NUCLEUS
A TRIP INTO THE HEART OF MATTER

RAY MACKINTOSH, JIM AL-KHALILI, BJÖRN JONSON AND TERESA PEÑA

Foreword by
Ben Mottelson

The Johns Hopkins University Press
Baltimore, Maryland

Introduction
Our world and its infinite variety

Sun, sky, trees, the hum of insects, birds swooping high above – we live in a beautiful world of amazing variety. Yet our world has another kind of beauty: the beauty that comes from an extraordinary unity lying behind the kaleidoscope of images. This 'unity' is quite straightforward, but it has taken thousands of years of human endeavour to uncover.

This book is about the human journey to the very heart of matter. The idea that everything in the world is made from the same fundamental ingredients goes back a very long way. However, it has taken centuries to work out just what those fundamental ingredients are. We now know that all things around us, this book, the air we breathe, the stars at night, are made up of just a hundred or so different kinds of atoms. The variety comes about because the atoms combine in more ways than we would ever be able to count or even imagine.

It was the ancient Greek scientists Leucippus and Democritus who first conceived of atoms. The fact that water could freeze or turn to steam and then turn back into liquid again, struck them as deeply significant. What was it that stayed the same during these transformations? Their inspired guess was that everything was made of tiny, indivisible objects that kept their identity through transformations like melting and boiling. While we now know that atoms can be 'split', this process is rather extreme and is not done merely by heating mixtures of chemicals together. Leucippus and Democritus certainly had the right idea.

Our picture of atoms has changed since ancient times – their properties are not just a matter of colour or shape. We do not think of acids as being composed of sharp atoms, or copper made of reddish atoms, for example. Different types of atoms are associated with different substances; there are atoms of gold, atoms of carbon, and atoms of iron, for example, but there are no atoms of ice cream. Whereas some substances are composed purely of one particular type of atom, others, like ice cream, are more complex and are built from different varieties.

Empedocles, a Greek living in Sicily around 450 BC, put forward the idea that everything is made up from just four elements: earth, air, fire and water. This basic idea was a very good one, and certainly an advance on the suggestion of one of his predecessors, Thales, who proposed that everything was made of water. But even Thales was on the right track in looking for a fundamental ingredient of everything he could see. After millennia of hard work, the number of elements identified has risen to a little over a hundred. Empedocles' earth, air, fire and water are not among them, but he had made a step along the right path. We now recognize that everything is made up of different combinations of elements, some familiar such as carbon, iron, hydrogen, oxygen; some exotic and, in some cases, extremely rare such as lutetium.

The Sun: powered by nuclear fusion.
(Courtesy SOHO consortium,
a project of ESA and NASA.)

JET: harnessing nuclear fusion on Earth. (Photograph courtesy of EFDA–JET.)

There are many questions to answer. Where do these elements come from? Why are the proportions what they are? Why, for example is there much more carbon than gold? The world might be prettier if gold predominated, but there would be no one to appreciate the glitter since carbon is the basis of life. Many enthusiastic searchers for extraterrestrial life concede that it is unlikely any form of life exists in the Universe that is not based on carbon. In addition, carbon is not the only element that is essential to life; most life on Earth also needs hydrogen, oxygen, nitrogen and a dozen or so other elements. Our existence depends on them being in sufficient supply. So how did all these vital elements get here? The answer to this question is one of the threads running through this book.

Another puzzle is why atoms have the individual properties that they do. We probably would not exist if carbon atoms were as heavy as lead atoms, so why is it that they are so much lighter? This will be explained along with why there are only about a hundred elements and not a million, or just four as Empedocles supposed.

In addition to investigating the composition of matter on Earth, we look out into the cosmos. Our nearest star, the Sun, is indispensable for life on Earth. All our energy ultimately comes from the Sun: the energy plants need for growing, the energy (occasionally released in spectacular lightning bolts) that drives the motion of the atmosphere, the energy that fuels all aspects of our modern age of technology.

Around the beginning of the twentieth century it was discovered that the Solar System was billions of years old. The origin of the Sun's energy then became an even greater mystery. What possible energy source could keep the Sun lit up for billions of years? No known source of heat came near to answering this. One hundred years later, that question can be answered. Why and how the Sun and the stars shine is another theme of this book.

The Crab Nebula, supernova remnant – most of the elements we know today were created in massive stellar explosions. (Courtesy European Southern Observatories.)

To answer all the above questions we need to understand the atomic nucleus, the tiny core of every atom. The composition of nuclei, and how and where they were created explains the nature of the fundamental elements, while the manner in which energy is released in their transformations is not only the story of how the Sun warms us, but that of the entire life cycle of stars.

Perhaps the best known facts about atomic nuclei are that they are involved in horrifying bombs and in what is seen as an unfriendly way of making electricity. These issues are addressed too, and we attempt to explain the true nature of radioactivity, a natural nuclear process. Applications of nuclear processes, including radioactivity, are embedded in many aspects of our lives, from medicine to geology, from testing jet engines to smoke detectors. To take one example, when employees at a nuclear plant in Canada threatened to go on strike, it emerged that 47,000 medical procedures per day would have been imperilled by the resulting shortage of radioactive isotopes.

Finally, how is it that we can be so confident of the age of the Solar System? This is an example of how radioactivity has helped solve a riddle in an entirely different field. The answer was obtained almost one hundred years ago by Ernest Rutherford through his discoveries about the natural nuclear processes that occur in rocks. To this day, the same ideas play a key role in unravelling the history of our planet Earth.

The attempt to explain everything in terms of a single substance which was started by Thales has now reached fruition. Today, we can describe the structure of all atomic nuclei in terms of an even smaller number of elementary particles.

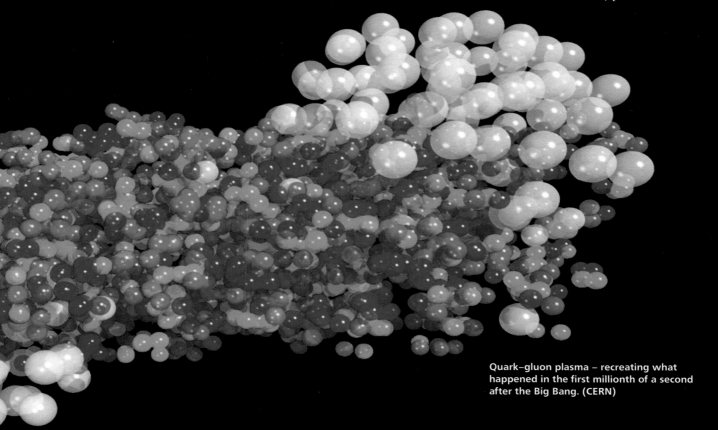

Quark–gluon plasma – recreating what happened in the first millionth of a second after the Big Bang. (CERN)

1. The size of things

From the immensity of space to the invisible world of the nucleus

Not so many generations ago, our ancestors would very likely have expressed the smallness of something by comparing it with a speck of dust floating in a shaft of sunlight. If they wanted to convey vastness, they might have made a comparison with a mountain, or even, if they were a little more sophisticated, with the Earth itself. If they were very learned, they might even have conjured up something a little larger: the sphere on which the 'fixed stars' were attached.

Today, our sense of the large and the small has expanded far beyond the comprehension of our forebears. This is due, at least in part, to the imaginative use of pieces of curved glass: the invention of the microscope and telescope. Our modern

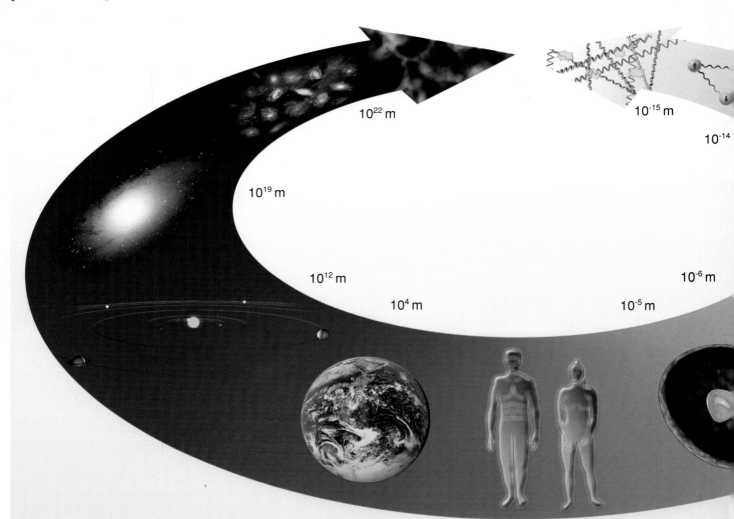

10^{22} m

10^{-15} m

10^{-14}

10^{19} m

10^{12} m

10^{-6} m

10^4 m

10^{-5} m

People stand midway between the immensity of the Universe and the unimaginable smallness of atomic nuclei and their building blocks. Yet what happened long ago involving these tiniest of objects set the entire Universe on its course. Today, we cannot understand the Universe as a whole without understanding Nature on the smallest scale. That is why the largest and the smallest meet in this picture.

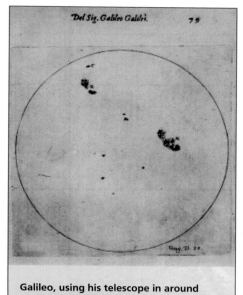

Galileo, using his telescope in around 1610, took the first steps beyond what we can see with our eyes alone. This is his own sketch of sunspots. (Courtesy Royal Astronomical Society.)

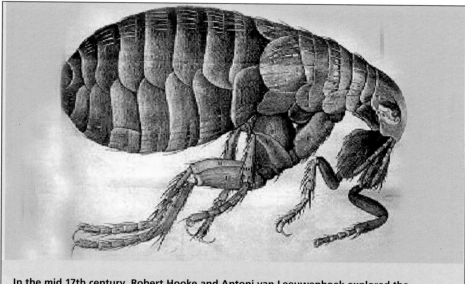

In the mid 17th century, Robert Hooke and Antoni van Leeuwenhoek explored the world of the very small. Their discoveries, like this flea drawn by Hooke, astonished their contemporaries.

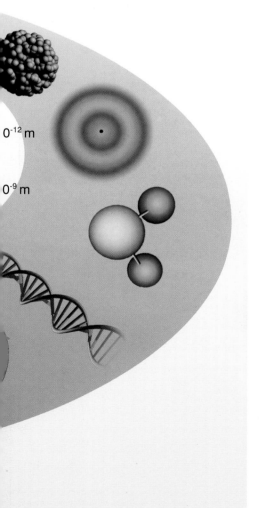

0^{-12} m

0^{-9} m

1 000 000 000 000 000 000 000 000	10^{24}	yotta	Y
1 000 000 000 000 000 000 000	10^{21}	zetta	Z
1 000 000 000 000 000 000	10^{18}	exa	E
1 000 000 000 000 000	10^{15}	peta	P
1 000 000 000 000	10^{12}	tera	T
1 000 000 000	10^{9}	giga	G
1 000 000	10^{6}	mega	M
1 000	10^{3}	kilo	k
1	**10^{0}**		
0.001	10^{-3}	milli	m
0.000 001	10^{-6}	micro	μ
0.000 000 001	10^{-9}	nano	n
0.000 000 000 001	10^{-12}	pico	p
0.000 000 000 000 001	10^{-15}	femto	f
0.000 000 000 000 000 001	10^{-18}	atto	a
0.000 000 000 000 000 000 001	10^{-21}	zepto	z
0.000 000 000 000 000 000 000 001	10^{-24}	yocto	y

The familiar 'kilometre' for thousand metres or 'millimetre', for a thousandth of a metre, do not help us much to talk about the really large or the very minute. This table presents the whole range of prefixes which apply not just to length (kilometre) but also to mass (kilogram) and other quantities.

understanding of the Universe involves both the unimaginably large and the unimaginably small. The nucleus is minuscule, but understanding its properties helps us to comprehend some of the ideas about matter on the greatest possible scale: the Universe itself.

Towards infinity

We are all familiar with objects on a human scale, but gaining a feel for distances much larger than the size of the Earth can be difficult. The distance of the Earth from the Sun, for example, is already quite hard to grasp. It is about 150,000,000,000 metres (93 million miles); if you could fly it in Concorde, the flight time would be about seven years – definitely not good for the circulation in your legs.

Working with such large numbers can be quite unwieldy and 150,000,000,000 is more usually written as 1.5×10^{11}. Moving out into the Universe introduces even greater quantities and astronomers use a much larger unit to measure length, based on the speed of light, which is 299,792,458 metres per second (186,280 miles per second). The distance light can travel in a certain time provides a convenient unit of length. It takes light about eight minutes to reach us from the Sun, so we say that the Sun is eight 'light-minutes' away. The furthest planet from the Sun, Pluto, is about six light-hours away, but it is the light-year that is most useful for expressing the enormous distances to objects outside our solar system. Since the light from the nearest star beyond the solar system, Proxima Centauri, takes four years to reach us, we say that it is four light-years away. This is about 38,000,000,000,000,000 or 3.8×10^{16} metres (733 million million miles).

On a larger scale, the Sun is one of about a hundred billion (a billion is a thousand million, 10^9) stars in a gigantic rotating disk called the Milky Way Galaxy, which is a hundred thousand light-years across. This means that light could have crossed it about 650 times since the dinosaurs became extinct.

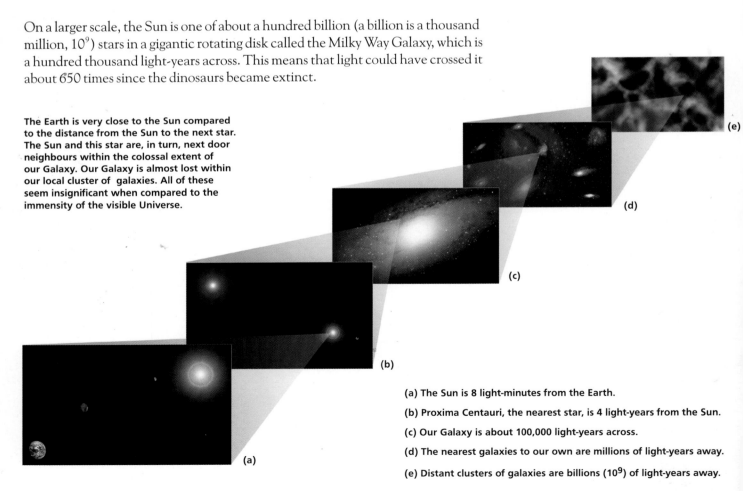

The Earth is very close to the Sun compared to the distance from the Sun to the next star. The Sun and this star are, in turn, next door neighbours within the colossal extent of our Galaxy. Our Galaxy is almost lost within our local cluster of galaxies. All of these seem insignificant when compared to the immensity of the visible Universe.

(a) The Sun is 8 light-minutes from the Earth.

(b) Proxima Centauri, the nearest star, is 4 light-years from the Sun.

(c) Our Galaxy is about 100,000 light-years across.

(d) The nearest galaxies to our own are millions of light-years away.

(e) Distant clusters of galaxies are billions (10^9) of light-years away.

On an even larger scale, space is studded with galaxies. They are grouped into clusters, separated by vast empty regions. The nearest large galaxy outside our own is Andromeda, visible to the unaided eye on a very clear, dark night as a faint glow the size of the Moon. Andromeda is about two million light-years away and is a member of the Local Group, a small cluster of galaxies which includes the Milky Way.

With a very powerful telescope, a vast number of galaxies can be observed, stretching out to such distances that the light coming from the very furthest would have been travelling for most of the time since the Universe was born. That was about 15 billion years ago and so they are the better part of 15 billion light-years away. The furthest point in space that we can ever hope to see marks the limits of what is called the observable Universe.

The size of the Universe is a measure of matter on the greatest possible scale. We will now set off in the opposite direction towards the smallest.

The galaxy NGC 2997; our own galaxy would look something like this from outside, a view we shall never have. (Courtesy European Southern Observatories.)

Under the microscope

The smallest unit of length with which most of us are familiar, is the millimetre (a thousandth of a metre), but for many objects which are only visible through a microscope, length scales are more sensibly expressed in terms of microns. A micron is a millionth of a metre, or a thousandth of a millimetre. A red blood cell inside your body is about seven microns in diameter; and a typical bacterium is only one micron across. Viruses, which are responsible for many diseases including the common cold, can range in size from a few tenths down to a hundredth of a micron.

Small as viruses are, they are not the smallest of objects; they are, in common with everything else, structures made of atoms. Everything is composed of a few types of atoms arranged in an uncountable number of different ways, and an atom is only about one ten-thousandth the size of a micron. Thus there are about a billion atoms in a typical virus and a single glass of water contains more atoms than there are glasses of water in all the seas and oceans on Earth.

This book is about entities much smaller even than atoms. Each atom has a nucleus at its centre that is typically about ten thousand times smaller still – a tiny bundle of concentrated mass and energy. The question of whether we have reached the limit of smallness with atomic nuclei is addressed later.

It is fascinating to explore how we can possibly know anything about things as tiny as nuclei. Microscopes have extended the range of our vision enormously, allowing bacteria to be observed, but no ordinary microscope will ever let us see a virus and they certainly will not work for atoms or nuclei. The reason for this is the nature of light.

Light is a form of electromagnetic radiation; the same type of phenomenon as radio waves, infrared, ultraviolet, X-rays and gamma rays. Electromagnetic radiation travels as a wave; the different types vary according to their wavelength, which depends upon the amount of energy the radiation possesses. It was not until the nineteenth century that Thomas Young was able to prove light has a wave motion.

The wavelength of visible light is about half a micron. Anything smaller, or any features on an object less than this amount, cannot be discerned using this type of electromagnetic radiation.

No optical microscope can distinguish objects smaller than the wavelength of the light shining on them, which is why viruses were not seen until the 1940s. Their detection had to await the invention of a wholly new kind of microscope which did not use visible light.

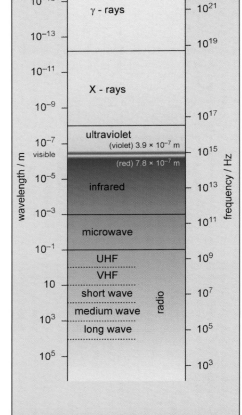

Visible light has wavelengths between ultraviolet and infrared. It is just one form of electromagnetic radiation. All the other forms are fundamentally the same, differing only in wavelength (metres) and frequency (Hertz, or cycles per second).

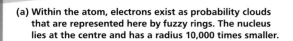

(d)

(c)

(b)

(a)

(a) Within the atom, electrons exist as probability clouds that are represented here by fuzzy rings. The nucleus lies at the centre and has a radius 10,000 times smaller.

(b) It is made up of protons (red) and neutrons (blue).

(c) They are themselves made of three quarks (two 'up' and one 'down' for the proton), which cannot exist by themselves.

(d) Much more speculative are the 'superstrings' or 'branes'. Quarks, electrons and other supposedly point particles might be unimaginably small strings.

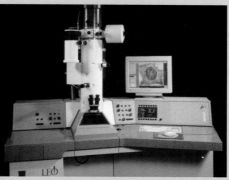
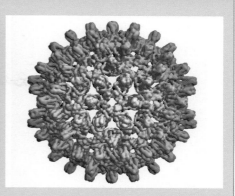

Here is a modern electron microscope flanked by an image of a human hair and, on the right and some 10,000 times smaller, a hepatitis B virus, showing its outer protein capsule which both protects it and provides the key which unlocks the defences of liver cells. (Courtesy LEO microscopes/John Berriman/MRC Cambridge.)

The revolutionary new instrument, an electron microscope, uses electron waves which have a much smaller wavelength. Electrons are the particles of electric charge that are present in every atom and which carry electric currents, but they can also behave as though they are spread-out waves. The wave character of electrons was the great discovery of a French prince Louis duc de Broglie. This discovery was a key step towards the understanding of the structure and properties of the subatomic world.

It sounds contradictory to describe electrons both as particles and as behaving like waves. This is the language of quantum mechanics, the physical theory in which electrons sometimes behave like waves and sometimes like particles. It is also the theory of the world of atoms and nuclei. For many purposes, electrons can be considered to be waves with a well-defined wavelength and this wavelength becomes shorter as the energy of the electrons is increased. Being waves, and also electrically charged, magnetic lenses can be made for beams of electrons and this allows the construction of electron microscopes. The limit of resolution is the same as for ordinary microscopes: the wavelength of the beam, but electrons have much shorter wavelengths than ordinary light. The shorter the wavelength of the electrons, the finer the detail that can be revealed.

In order to see smaller and smaller things we use electrons with higher and higher energies because higher energies mean shorter wavelengths. As electrons are charged particles, they will feel a force if put into a vacuum with an electrical voltage across it. This force causes them to accelerate. This is the manner in which electrons are accelerated in ordinary television tubes or computer monitors. In such applications, the electrons are focused to a point which moves over the screen, tracing out the picture. Electron microscopes generally accelerate electrons to energies ten or more times higher than in a television tube in order to get the wavelength small enough to see viruses.

Atoms are much smaller still than viruses. Today it is understood how they are arranged in most materials, but as late as the first decade of the twentieth century the very existence of atoms was seriously doubted by many scientists. It is truly remarkable that we are able to learn anything at all about things so small.

Louis Victor Pierre Raymond duc de Broglie, 1892–1987, the first person to glimpse the wave nature of matter. (Copyright the Nobel Foundation.)

Imaging sheets of atoms: this is the surface of a crystal of iron silicide as seen using a device called a scanning tunnelling microscope. The smallest step corresponds to a layer one atom thick. (Courtesy Nanoscale Science Laboratory, Cambridge.)

How do we know that light is waves?

When light falls on a barrier with tiny holes or slits in it, a pattern of light and dark can be found on a screen placed on the far side. The same sort of thing happens with any waves, including water waves where it is easier to see what is happening.

Waves spread out when they encounter objects, something most easily observed when their wavelength is comparable to the size of the object. This spreading out is known as diffraction. Waves from different holes interfere with each other and where a trough arriving from one hole meets a crest from another they can cancel each other out. For light waves this means there are dark patches at such places, as Thomas Young discovered.

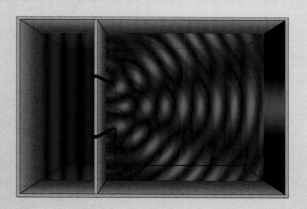

Another type of electromagnetic radiation is X-rays. These have wavelengths typically a thousand times smaller than visible light. Unfortunately it is not possible to build X-ray microscopes because there is no suitable system of lenses that can bring X-rays to a focus within such an instrument. Nevertheless, almost as soon as they were discovered, X-rays were used to study the structure of matter. The secret is to use a universal property of all waves: diffraction.

Early in the nineteenth century Thomas Young demonstrated how a grating creates an interference pattern using visible light. Working backwards from complicated diffraction patterns allows scientists to determine what sort of 'grating' gave rise to those particular patterns, and atoms in a crystal are, in effect, a grating. This method was used to determine, for example, how far apart the sodium and chlorine ions are in a crystal of common salt. Much more interesting crystals than common salt have been studied with X-ray diffraction. It was by looking at the way X-rays were diffracted by DNA that Francis Crick and James Watson unravelled the double helix structure of this key molecule of life.

The method of diffraction, applied to electron waves rather than X-rays, allows the size and shape of atomic nuclei to be measured. Real images of atoms in the surface of a crystal can be obtained, taken using a particular type of electron microscope called a scanning tunnelling microscope.

We now leave the realm of atoms for the realm of nuclei, ten thousand times smaller. There was no reason to suspect that Nature had structures on such a small scale. Clues of this 'sub'-atomic world came in the shape of a wholly unexpected new phenomenon: radioactivity.

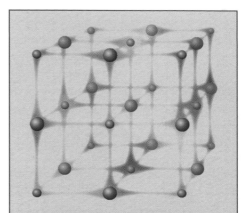

A crystal of common salt consists of alternating sodium and chlorine atoms arranged in a regular cubic array. The spacing between the atoms can be worked out from the way X-rays diffract from the crystal.

Thomas Young, 1773–1829, polymath. It was Young who first provided convincing evidence of the wave nature of light. (AIP Emilio Segrè Visual Archives.)

2. The discovery of nuclei

How a new world was revealed by radioactivity

As the nineteenth century drew to a close, many scientists felt that all the important questions in science had been answered. This was wishful thinking. Scientists could not even agree that matter was made of atoms let alone explain something as basic as why copper is red! The majority just thought that it was because it had 'redness'. Today, such a statement would not be acceptable; the colour of copper can be explained with far more convincing arguments, but only because the structure of atoms is understood more thoroughly.

Gaining the means to investigate the structure of something as small as an atom began in 1895 when Wilhelm Röntgen astonished the world with the discovery of X-rays. He had been investigating the source that had fogged some photographic plates which had been kept carefully wrapped in his laboratory. Röntgen eventually tracked the source to a gas-discharge tube. The discovery of X-rays sent shock waves around the world. Here were mysterious new rays that could pass through solid materials to cloud photographic film. Very soon, people all over the world were marvelling at photographs showing the bones in their hands.

The mysterious X-rays seemed to be streaming out of a part of the tube that glowed in the dark; it was fluorescent. This property was investigated by Henri Becquerel, who studied a uranium compound he knew emitted this eerie glow. Like X-rays, he found the radiation from uranium could pass through the black paper protecting photographic film from light, and leave its traces on the undeveloped film. Unlike X-rays, which disappeared when the X-ray tube was switched off, the radiation from the uranium was emitted continuously. Becquerel soon found that any compound of uranium, even pure uranium metal, worked just as well.

Wilhelm Röntgen, 1845–1923, whose accidental discovery of X-rays in 1895 opened a new window on the world. (Copyright the Nobel Foundation.)

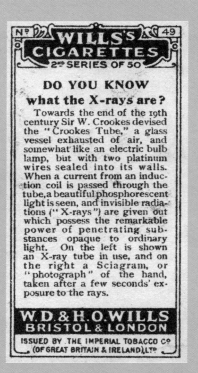

WILLS'S CIGARETTES
2ND SERIES OF 50

DO YOU KNOW what the X-rays are?

Towards the end of the 19th century Sir W. Crookes devised the "Crookes Tube," a glass vessel exhausted of air, and somewhat like an electric bulb lamp, but with two platinum wires sealed into its walls. When a current from an induction coil is passed through the tube, a beautiful phosphorescent light is seen, and invisible radiations ("X-rays") are given out which possess the remarkable power of penetrating substances opaque to ordinary light. On the left is shown an X-ray tube in use, and on the right a Sciagram, or "photograph" of the hand, taken after a few seconds' exposure to the rays.

W. D. & H. O. WILLS
BRISTOL & LONDON
ISSUED BY THE IMPERIAL TOBACCO CO (OF GREAT BRITAIN & IRELAND) LTD.

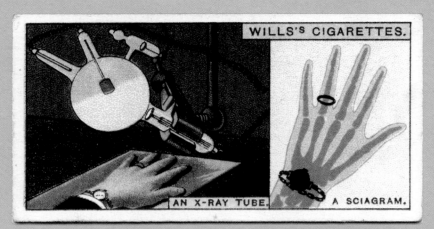

WILLS'S CIGARETTES.
AN X-RAY TUBE.   A SCIAGRAM.

X-rays stimulated scientists around the world into action. They also aroused a great deal of public interest which had not abated decades later, as the cigarette card shows.

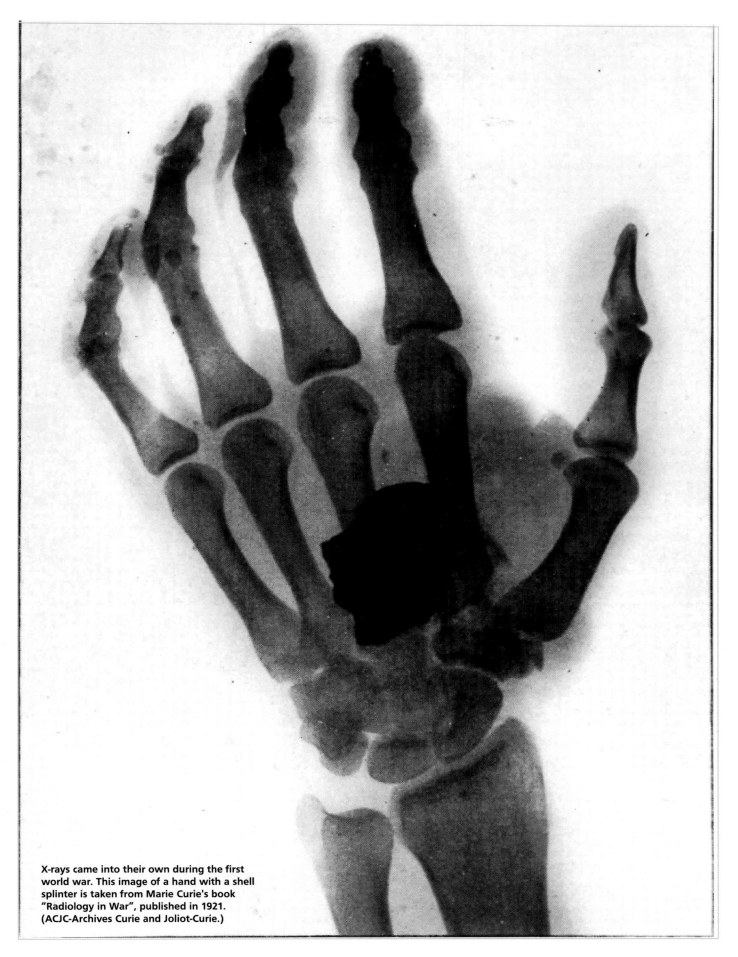

X-rays came into their own during the first world war. This image of a hand with a shell splinter is taken from Marie Curie's book "Radiology in War", published in 1921. (ACJC-Archives Curie and Joliot-Curie.)

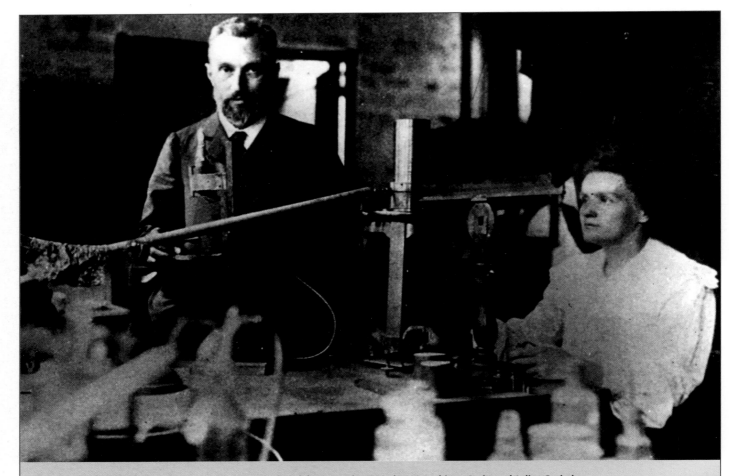

Pierre, 1859–1906, and Marie Curie, 1867–1934, in their laboratory in 1898. (ACJC-Archives Curie and Joliot-Curie.)

For a while, only uranium and its compounds were known to emit these new rays. Further studies by Marie Curie in Paris, however, revealed that another element, thorium, had the same property. Again, the chemical form of thorium did not matter, all forms – thorium metal, thorium oxide etc. – emitted exactly the same rays, which, just as for uranium, poured out steadily without any external source of energy being applied.

Until the 1890s, thorium was just an element whose oxide was used to make gas lighting mantles bright, while uranium had been used since Roman times as a yellow pigment in glass. These rather boring elements were about to leap to the centre of the scientific stage and offer unique clues about the structure of matter.

The hunt for radioactive elements

Marie Curie was one of the few scientists of the time who was inspired to continue Becquerel's work. She was the one who first used the word 'radioactivity' to describe the radiation emitted by uranium and thorium. Becquerel's work took some time to gain significance, perhaps because of the recent discovery of X-rays which had very obvious applications in medicine and industry. X-rays had gripped both the public and scientific imaginations; popular cartoons made jokes about how the new rays would penetrate the many layers of Victorian petticoats. However, Marie Curie had the instinct of a great scientist to spot a wholly new window into nature.

Lord Rutherford is celebrated on the New Zealand 100 dollar bill.

Joseph John Thomson, 1856 –1940, discoverer of the electron, the first fundamental particle of Nature to be identified. (Copyright the Nobel Foundation.)

Born Marie Sklodowska, she had recently arrived in Paris from Poland and married Pierre Curie, who had himself already made some important discoveries in physics. The story of how Marie Curie discovered polonium, a new radioactive element lurking in trace amounts in uranium ore, followed by radium, an element far more radioactive than uranium and very much rarer, is an inspiring one.

Unlike uranium, pure radium glows in the dark. It also gives out heat. Even the tiny amounts that Marie Curie was able to extract radiated a measurable supply of heat. Radium soon became highly prized as a source of powerful radiation for further experiments. From the earliest days, it was also studied as a possible cure for cancer. This was the beginning of radiotherapy.

Rutherford joins the quest

Meanwhile, a young New Zealander called Ernest Rutherford had arrived in Britain to work in Cambridge with J.J. Thomson, the greatest experimental physicist of the day. 'J.J.' was interested in how X-rays made air conduct electricity. If you rub a plastic ruler on a woollen sweater, it will pick up small pieces of paper. This is because it acquires an electric charge on its surface. Near an X-ray machine, such a ruler quickly loses its charge because the air becomes conducting and the charge drains away. The reason for this is that the X-rays ionize the air by freeing some electric charge (electrons) from their atoms.

Thomson and Rutherford initially set out to explore this ionization by X-rays. It was known that Becquerel's new kind of radiation also ionized air and exploring this new phenomenon of radioactivity soon became Rutherford's passion.

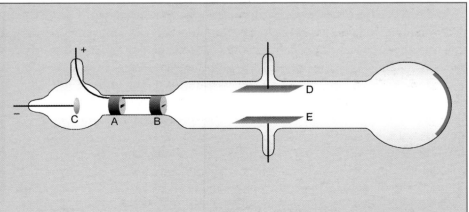

J.J. discovered the electron with this apparatus, a forerunner of the cathode ray tube found in computer monitors and TV sets. Rays emitted from the 'cathode', C, strike the fluorescent screen at the far end of the tube, leaving a spot of light. Thomson's genius was to interpret the paths of the rays in terms of charged corpuscles, later called electrons. Electrodes A and B were at a positive voltage, drawing off the electrons through the slits. These were then bent by electric and magnetic fields. Electrodes D and E supplied the electric fields, external coils provided the magnetic fields.

At about this time, temporarily leaving his work on X-rays, Thomson made the momentous discovery that electric charge exists in packets or particles of electricity, now called electrons. For this, he won the Nobel Prize. It is now known that electric currents in wires are just the flow of these electrons; and electrons are a key component of every atom.

Shaking the foundations of science

While Thomson worked on the nature of electric charge, Rutherford had made his first big breakthrough. He had found that there were two distinct kinds of radioactivity. This was not a trivial discovery in the days before modern instruments; it had the hallmarks of insight and ingenuity that were to make Rutherford famous.

He discovered that one kind of radiation was highly ionizing and very easily absorbed, even by thin sheets of paper. He called this 'alpha' radiation. The other type was far more penetrating and was named 'beta' radiation. Soon afterwards, Rutherford established one of the most puzzling and characteristic features of radioactivity: no matter how much radioactive material is present initially, half of it will disappear after a time interval known as the 'half-life'. After two half-lives, only a quarter of the original material remains, and so on. This property, known as exponential decay, is tied up with the quantum mechanical nature of atoms described in the next chapter. Rutherford followed these findings with another important discovery that earned him, somewhat to his amusement, the Nobel Prize in chemistry.

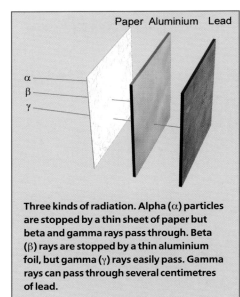

Three kinds of radiation. Alpha (α) particles are stopped by a thin sheet of paper but beta and gamma rays pass through. Beta (β) rays are stopped by a thin aluminium foil, but gamma (γ) rays easily pass. Gamma rays can pass through several centimetres of lead.

The end of chemistry's most cherished belief

Elements are the foundation stones of chemistry. The vast variety of substances around us are made from combining elements – such as carbon, oxygen, hydrogen and nitrogen – in different ways. At the turn of the twentieth century, a firmly held belief was that atoms never change. Carbon atoms remain carbon atoms; iron stays iron; gold cannot be made from lead; and once uranium, always uranium. Rutherford and his colleague Frederick Soddy, who was at that time working with him at McGill University in Montreal, Canada, overturned this idea. They found radioactivity was a process that could transform one element into another. Somehow, one type of atom emitted a particle and became another type of atom. No doubt mindful of the ill-repute arising from thousands of years of futile attempts to make gold from lead, Rutherford is reported to have said to his friend: "Don't call it transmutation, Soddy, or they'll have our heads as alchemists!"

The idea of atoms transmuting from one type to another met with initial resistance, even from the Curies. Within a few years, however, such overwhelming evidence was presented by Rutherford, Soddy and many others, that transmutation became widely accepted. A great number of people were soon involved in unravelling the complicated story of which element transmuted into what. This task was made much easier by knowing the nature of alpha and beta radiation, as well as by the discovery of a third type of radioactivity called gamma rays.

The nature of radiation

Following the discovery of alpha and beta radiation, physicists were keen to investigate their nature. Beta particles were relatively easy to identify. Their paths are deflected easily by strong magnets which shows they are electrically charged and are thousands of times lighter than atoms. Beta particles were soon identified as electrons – the particles recently discovered by Thomson to be the smallest carriers of charge.

The nature of alpha particles was harder to establish. They are certainly electrically charged, but are harder to deflect by magnets, which means that they have to be much heavier than electrons. Rutherford discovered that alpha particles are helium ions: helium atoms that have been stripped of their electrons, leaving them positively charged.

Age of the Earth mystery

The idea of transmutation, together with an understanding of what alpha and beta rays are, opened up the possibility of using radioactivity for a variety of measurements. By 1905, Rutherford could claim to know the age of the rocks he was carrying in his pocket simply because he knew that alpha particles were helium ions.

Rocks naturally contain a small amount of radioactive uranium that transforms into another type of element, spitting out helium ions in the process. These ions soon pick up their missing electrons and become neutral helium atoms which are trapped in the rocks. These minerals are studied in a laboratory to see how much helium they hold.

Knowing the amount of helium locked away in the rocks, Rutherford could determine how much time had passed since the rocks were formed. He had to take into account the amount of uranium in the rock along with other elements, such as lead, which are the end products of a whole chain of transmutations starting with uranium. Today, analyses like this are crucial to geology, and dating rocks by studying the decay of radioactive minerals is a highly developed application of nuclear techniques.

Having demolished one of chemistry's most cherished beliefs – namely that the elements cannot change from one type to another – radioactivity was also beginning to threaten a cornerstone of physics. This was the law proclaiming that energy is always conserved. Although it can be transformed from one form to another, the total amount of energy of a system always remains the same. How could radioactive materials seemingly pour out energy from nowhere? This was a real dilemma. Later it was found that the total mass of nuclear particles becomes less following radioactive decay. It was Albert Einstein who first showed that a loss of mass means a release of energy.

But in 1904, before Einstein's theory of relativity, nuclear energy solved another vexing problem. The great physicist Lord Kelvin had estimated that the Earth could not be more than 100 million years old, otherwise its interior would have had time to cool down. Geologists argued that it was much older; and biologists knew that life would have needed much more time to evolve. Rutherford pointed out that the radioactivity of minerals in the ground was an additional source of heat, which made a much older Earth possible. In fact, the rocks in his pocket were themselves much older than Kelvin's estimate of the age of the Earth.

The next application of radioactivity was momentous: studying the inner structure of matter itself. The result was the discovery of the atomic nucleus. The physicist in the vanguard of this research was a New Zealander working at Manchester University in the North of England – Ernest Rutherford.

Our blue planet is approximately 4.5 billion years old. If it were not for the nuclear decay processes at the core, the Earth would have cooled down to a solid ball long ago, without plate tectonics, earthquakes or volcanoes. (Courtesy of M. Jentoft-Nilsen, F. Hasler, D. Chesters NASA/Goddard and T. Nielsen University of Hawaii.)

A Solar System analogy for the strong deflection of alpha particles passing through an atom. A comet's path is very curved close to the Sun. This is because the Sun's gravitational force is large near the Sun and much smaller further away. If the mass of the Sun were spread through the huge volume of the Solar System, a body coming from outside the Solar System would only be weakly deflected. In the same way, an alpha particle would be weakly deflected if the positive charge were not concentrated in a tiny volume, the nucleus. (Julian Baum)

Geiger and Marsden bombarded a whole range of substances with alpha particles. Counting individual zinc sulphide scintillations in the dark using a microscope, they then painstakingly measured what proportion of the particles went straight through various thin metal foils, what proportion were bounced back and the numbers scattered at all the intermediate angles. Extreme patience was needed to get a reasonable number of counts for the large angles to make the results meaningful. Rutherford freely admitted that he did not have the patience for the job.

Geiger and Marsden's measurements matched Rutherford's predictions perfectly. These pioneers with their acute eyes and patience were rewarded with a momentous discovery: atoms have nuclei. Their discovery revolutionized not just physics, but also astronomy, chemistry, biology and our everyday view of the world.

The experiments even made it possible to say that the hydrogen and helium nuclei must be less than four femtometres in diameter (a femtometre being 10^{-15} of a metre) and that the gold nucleus must be no more than ten times this size. To put these numbers into perspective, one femtometre is to one micron what a micron is to a kilometre. Or to put it another way, the size of a typical nucleus relative to one metre is about the same as the size of a pinhead relative to the distance between the Earth and the Sun.

These experiments did much more than prove the existence of atomic nuclei. They inaugurated a whole new method of studying nature: the scattering experiment. Even today, the most important way to study atoms and their nuclei is to fire a beam of particles at a target and watch how they are scattered.

It is certainly not at all obvious that we can learn anything about atoms that are 10^{-10} metres across, or nuclei that are between 10,000 to 100,000 times smaller still, but, nearly a hundred years later, we have learnt a huge amount, mostly from scattering experiments. Today, such experiments are vastly more complex than Geiger and Marsden's, but they are based on the same principle. Particle accelerators can provide particles at far higher energies and intensities than radioactive sources, and scientists no longer sit in darkened rooms counting flashes of light with their eyes. Instead they use complex electronic detectors and an array of computers to sort and store many gigabytes of data from the experiments.

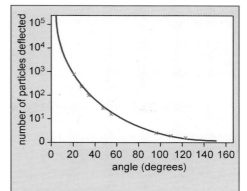

The solid line shows Rutherford's prediction that many alpha particles would be scattered through small angles, but much fewer at large angles. The dots show the measurements of Geiger and Marsden agreeing very well. According to pre-nuclear models, all alpha particles would hardly be deflected at all.

The apparatus used by Geiger and Marsden to verify Rutherford's prediction of alpha particle scattering. R is the alpha emitter, F is the foil, S is the fluorescent screen and M is the viewing microscope. Screen and microscope could be moved around so that the number of alpha particles scattered at different angles could be counted.

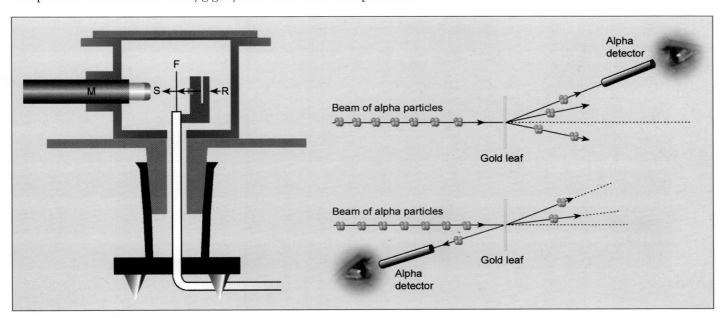

Rutherford knew that his nuclear model left many unresolved problems. For example, if all the positive charge was in the tiny central nucleus while the negatively charged electrons orbited around like planets around the Sun, why did these electrons, which are attracted to the positive charge, simply not fall into the nucleus and neutralize it? There is an important difference between an atom and the Solar System. The Earth and other planets are all in stable orbits around the Sun, whereas the orbiting electrons should be losing energy as they go round the nucleus. This is because electric charges emit radiation when accelerated; a particle moving in circular motion is accelerating, even if its angular velocity is constant. This loss of energy means that the electrons should rapidly spiral in towards the nucleus. The resolution of this puzzle would require a stroke of genius from a young newcomer.

Electrons in orbit

A young Dane called Niels Bohr was on an extended visit to Rutherford in Manchester. Bohr's answer to Rutherford's problem also explained a puzzling property of the light given out when an electric current is passed through hydrogen gas. A Swiss schoolteacher, Johann Balmer, had earlier discovered a remarkable mathematical pattern when analysing the colour of this light. Scientists believed that a strikingly regular pattern like the so-called 'Balmer' series should have a clear-cut explanation. Bohr solved the mystery.

Bohr's models of hydrogen (top left), helium (top right) and neon atoms.

The basic idea of his model was that the electrons do not fall into the nucleus because they revolve around it in fixed orbits without radiating any energy.

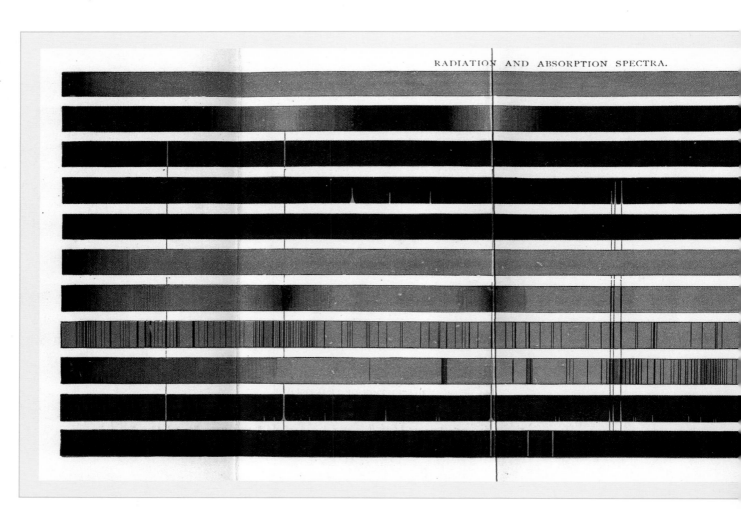

RADIATION AND ABSORPTION SPECTRA.

In addition, Bohr's model only allowed certain distinct orbits, almost as if there were a number of invisible rail-tracks around the nucleus on which the electrons had to run. These paths were determined by the standard theories of physics of Bohr's day, with the crucial addition of a certain new ingredient: the quantum theory proposed by Max Planck and Albert Einstein.

Planck and Einstein had used the revolutionary idea that energy comes in packets called 'quanta' to explain a number of puzzling properties of radiation and heat. In 1913, Bohr applied these ideas to atoms and gave a first explanation of why electrons in atoms did not fall into the nuclei. His model convinced many people of Rutherford's nuclear picture of the atom, and was a key step towards the quantum theory that was to become the basic theory of all phenomena at the atomic scale.

That was far from the end of the story however. While most people were convinced that there must be something to Bohr's model, it failed to account for many of the properties of atoms other than hydrogen. It was not until the development of quantum mechanics in the 1920s that the full picture of atomic structure emerged.

Ingredients to make an element

Despite its limitations, the broad concept of Bohr's atomic model remains applicable today. It furnishes a framework for understanding the chemical elements and also provides the language used to describe atoms and nuclei.

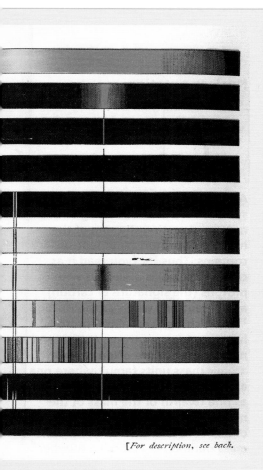

[For description, see back.

Each element has a unique light 'fingerprint' or spectrum. The different colours emitted when the element is heated can be separated by passing the light through a prism or other kind of 'spectroscope'. The resulting pattern is called an emission spectrum. If white light, composed of all colours, passes through a vapour containing atoms of a particular element, then those atoms all absorb light of the same colours. The result is called an absorption spectrum and can be seen as a series of black lines shown here. This picture, showing a range of different spectra, was the frontispiece of Norman Lockyer's "Elementary Lessons in Astronomy" 1874. Lockyer became famous for discovering helium on the Sun before it was known on Earth – and it is also for this reason that the element was named after the Greek word for the Sun, 'Helios'.

3. Particles or waves?

Strange laws at the heart of matter

Television news items or films sometimes show someone using a Geiger counter… maybe a prospector is searching for uranium, or perhaps a hospital worker is accounting for vital radioactive materials used to treat cancer. Geiger counters are particle detectors that make a characteristic clicking sound every time a high energy particle enters. Click – one particle; click – another particle.

It is known that X-rays are emitted by electrons in the atom, whereas gamma rays, electromagnetic radiation of an even higher energy, are emitted from within the atomic nucleus. The clicking of a Geiger counter is a sign of quantum weirdness – the gamma ray waves are behaving as particles. In the same way that particles like electrons can sometimes behave as waves (we have mentioned how the wave properties of electrons allow us to study the structure of matter), all electromagnetic waves can occasionally act as particles.

Every type of particle has a name, and particles of light are called photons. All electromagnetic waves, whether they are gamma rays, X-rays, visible, ultraviolet light, or radio waves, consist of photons. Another important property of electromagnetic waves is that, unlike other forms of waves such as sound, they do not need a medium to carry them. Sound waves require air to travel through, but all forms of light can propagate through empty space. It is because photons happily travel in the near vacuum of space that the radiation emitted by the Sun can heat the Earth.

J. L. Cassingham with one of his company's Geiger counters, featured in an advertisement from 1955. (Courtesy J. L. and Curtis Cassingham.)

The strange world of photons

Every day we use photons of visible light to find our way about and to enjoy our environment. We can extend our visual sense with microscopes to study objects that would otherwise be too small to see with the naked eye. But ordinary microscopes are no help when it comes to discerning structures smaller than the wavelength of visible light. To see the details of such structures we need to use photons with shorter wavelengths. Investigating the diffraction patterns caused by shining X-rays through salt crystals, for example, allows the spacing between the atoms to be measured. The process by which the diffraction pattern is created is a consequence of quantum effects.

Imagine turning down the intensity of the X-ray beam until just one photon passes through the crystal at a time. A single photon like this can act like a particle, blackening a single minute grain of sensitive material on a photographic film, but if we fire a huge number of photons one at a time at the crystal, aiming exactly the same way each time, then the spots eventually build up into a diffraction pattern.

While each photon arrives at the film as a tiny particle, the only way the diffraction pattern could build up is for each photon to 'see' the whole array of ions. Somehow, like a wave, it must pass through the whole crystal and yet, like a particle, it will blacken a single grain on a photographic film. A single minute spot on a film is not a diffraction pattern; only after many individual photons have passed through the crystal does the pattern build up.

Weird quantum effects

In our X-ray shoot-out, the photons do not pile up at the same spot on the film even though they were aimed and fired in exactly the same way. The pattern that builds up reveals the wave nature of the particles as more and more photons hit the film. It is impossible to predict where any particular photon will turn up on the film, yet the overall pattern that eventually builds up consists of groups of spots clustered together in bands with gaps in between where there are no spots. This interference pattern is a feature that always shows up whenever quantum effects are important.

Many people find this 'uncertainty' one of the most disconcerting characteristics of quantum behaviour. For centuries, science has been based on the idea that a particular situation always leads to the same outcome. In the X-ray diffraction experiment, this rule is broken: it is simply impossible to predict where a photon will leave its mark on the film.

All is not lost, however. A new, but limited, kind of predictability emerges from the ruins of the old predictability, in which people believed they could say where a particle would go. The pattern that builds up from a large number of unpredictable spots is itself predictable, and can be reproduced whenever the entire experiment is repeated. We just have to abandon the idea that we can ever say where an individual photon will blacken the film. This strange new interplay of predictability and unpredictability turns up repeatedly in the realm of atomic nuclei.

Further quantum weirdness

In diffraction experiments, the regular array of atoms in the crystal determines the final pattern. Somehow each photon has to 'see' the crystal as a whole – even if it turns up at a single point on the screen. Try to determine exactly where the photon went through the crystal, however, and the diffraction pattern is destroyed. It is as

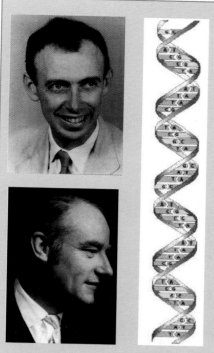

James Watson (top), 1928–, and Francis Crick, 1916–, analysed the diffraction of X-rays by DNA. In 1953 they discovered the remarkable double-helix structure of the molecule on which our genome is encoded. (AIP Emilio Segrè Visual Archives.)

Werner Heisenberg, 1901–1976, left, one of the creators of quantum theory, with his mentor Niels Bohr, 1885–1962, the first person to apply quantum ideas to atoms. Heisenberg was the first to incorporate neutrons into a model of nuclei. (AIP Emilio Segrè Visual Archives.)

if the photon fills a large volume; yet when it is detected, it is found at particular point. This is one consequence of Heisenberg's Uncertainty Principle – a law that affects absolutely everything in the sub-microscopic world.

Quantum effects are not just confined to photons – electrons, protons and neutrons behave strangely too, which is one reason why it is so hard to visualize what an atom really looks like.

Electrons orbit the nucleus, but not in a manner that resembles a miniature Solar System of planets orbiting a sun. If electrons behaved like planets, they would pass through a sequence of precise and measurable positions. The position of an electron can only be given as a probability; an electron does not have a precise position until we try to measure where it is. Moreover, just as trying to measure where the X-ray passed through the crystal destroys the diffraction pattern, so measuring the position of an electron will almost certainly knock it right out of the atom.

There is scarcely a more natural idea than that an object is located at a certain place; an idea that has risen from our experience of everyday objects. The new quantum view of the world forces us to rethink what is meant by 'position'. In the case of the electron, all we can do is predict the likelihood of the electron being found at a given place in an atom. This is not due to any shortcomings in our theories or measuring instruments, but rather because it is a property of nature itself at this minuscule scale.

The wave-like behaviour of electrons can also be used to measure the size of nuclei and the structure of crystals. In a similar manner to X-rays, predictable

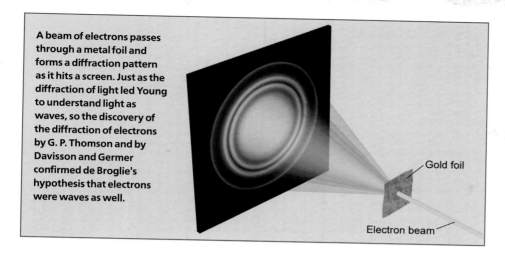

A beam of electrons passes through a metal foil and forms a diffraction pattern as it hits a screen. Just as the diffraction of light led Young to understand light as waves, so the discovery of the diffraction of electrons by G. P. Thomson and by Davisson and Germer confirmed de Broglie's hypothesis that electrons were waves as well.

Gold foil

Electron beam

and repeatable diffraction patterns can be built up spot by spot on a photographic film by using electron beams, although there is just no way of predicting where a particular electron will leave its mark.

Electrons and photons do not have a monopoly on quantum weirdness. All atomic and nuclear particles, including protons and neutrons, exhibit the same schizophrenic behaviour we call wave–particle duality. Sometimes they behave like a wave and show diffraction effects, and at other times they behave like particles.

Since 1925, physicists have come to accept that the world of atoms and nuclei is utterly different from the familiar world of objects measured in metres and centimetres. One of the pioneers of quantum theory, Niels Bohr, once said: "If you're not shocked by quantum mechanics, you don't understand it." Another legendary physicist, Richard Feynman, went further, claiming that no one understands quantum mechanics. Certainly, physicists who use quantum physics every day still find it deeply puzzling.

Living the half-life

Radioactivity provides a clear example of the concept of indeterminacy; that is, identical initial situations lead to different outcomes. Consider a million identical uranium nuclei, identical in a way which goes beyond the everyday meaning. They are not just identical in the sense that toothbrushes coming off a production line are the same, the nuclei are indistinguishable by any conceivable measurement. These nuclei are also unstable, and undergo radioactive decay, but they do not decay in an identical time. All that can be said is that they have an equal likelihood of decaying in a given interval of time.

After a certain time (the half-life), half of the nuclei will have decayed, but there is no way of predicting when a particular uranium nucleus will decay. However, one very interesting aspect of determinism does hold true: if we take another large sample of uranium nuclei then it will take the same length of time for half of it to decay as it took for the first sample.

We can imagine a way to avoid the indeterminism in the decay of a nucleus if the unpredictable nature of decay is only a matter of our ignorance. Perhaps all the seemingly identical uranium nuclei are not identical after all. Rather, they have some kinds of unobservable characteristics that distinguish them from each other. These 'hidden variables' would be locked away inside each individual uranium nucleus, telling it when to decay.

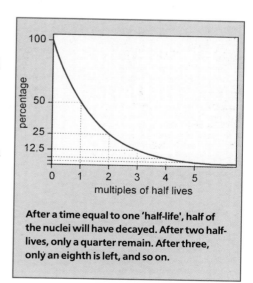

After a time equal to one 'half-life', half of the nuclei will have decayed. After two half-lives, only a quarter remain. After three, only an eighth is left, and so on.

However, most physicists are these days convinced that Nature does not resort to hidden variables to explain the unpredictable behaviour of radioactive decay. We just have to live with the idea that Nature is unpredictable.

Making tracks

'Taking a quantum leap' is an expression that has filtered into everyday language – but what does it mean? Quantum leaps are another way the quantum world differs from our familiar world of human-sized objects.

Once again the radioactive decay of nuclei provides a window into the subatomic domain. Visible trails left by particles emerging from nuclei can be imaged in cloud chambers. A great deal of information can be determined from the tracks, including the energy of the particle. The particles with the highest energy leave the longest trails: they travel the furthest before slowing down and coming to a halt. Close measurement of tracks in cloud chambers reveal some alpha particles possess a little bit less energy than others.

Measuring the energy of the alpha particles is very important, and cloud chambers have long since been replaced by silicon detectors. These are rather special

Each of these tracks is a trail of tiny water droplets left by a single alpha particle passing through a cloud chamber. These tracks are evidence of particles, but to understand how alpha particles are emitted from nuclei, they must be treated as waves.

silicon chips that give an electrical signal when a nuclear particle hits them. This signal allows the energy of the particle to be measured very accurately. Results from silicon detectors confirm that some alpha particles have a little less energy. The energies differ by distinct amounts that get progressively lower than the maximum energy.

An important physical concept that is employed here is the law of conservation of energy. Richard Feynman called it the cornerstone of physics. It means that the total energy of a system is always the same. If some alpha particles, emitted by exactly the same kind of nuclei, have a little less energy than the others, then the missing energy must stay in the so-called daughter nucleus: the nucleus left behind when the alpha particle is emitted.

The energy the alpha particle has as it flies out of a nucleus comes from nuclear energy that is normally locked away inside the nucleus. Since the alpha particles only ever come out with one of a set of distinct energies, then the daughter nucleus left behind can only have one of a set of certain values of energy. These values are known as energy levels and are often represented symbolically in diagrams by a series of horizontal lines. The fact that a nucleus can exist only in a series of distinct energies is an example of a new universal property. What is true of the daughter nucleus created in alpha decay is true of all nuclei, as well as of atoms and molecules. All these tiny objects can possess any one of a series of fixed amounts of energy, known as energy states. It is only when we compare this with everyday life that we realize just how strange it is.

Think of a car. The faster it moves, the more energy it has. In principle it can have any energy up to the energy corresponding to its top speed. Now imagine that the fuel runs out when the car reaches top speed. It slows down, passing continuously through all possible energies until it comes to a halt. It does not slow down in discrete steps and there are no missing energies that are avoided as the speed decreases.

Alternatively, imagine a red-hot poker cooling down. Since heat is a form of energy, the poker loses energy as it cools, but it does not cool down by stopping at a particular temperature and then jumping to a lower one. It cools continuously, passing through all possible temperatures on the way. This is so familiar that we never feel the need to call it a 'no-jump' process.

In the atomic world, nuclei do jump in energy. If a nucleus emits an alpha particle with less than the maximum possible energy, the daughter nucleus is left with a certain amount of extra energy above its 'ground state' (when it has the lowest energy possible). The daughter is said to be in an 'excited state'. So, unlike the car or the red-hot poker, it loses energy by jumping to a state with lower energy. It might do this several times until it eventually reaches the ground state.

With each jump, it gets rid of a packet of its energy. This energy packet is none other than a gamma ray photon. Most of the clicks a uranium prospector hears coming from a Geiger counter are caused by gamma ray emission, a secondary effect of the alpha decay process. Most alpha particles themselves do not penetrate through the walls of the Geiger counter.

The term 'quantum leap' refers to the way the nucleus loses energy by discrete amounts rather than smoothly, but the energy lost by a single nucleus is absolutely

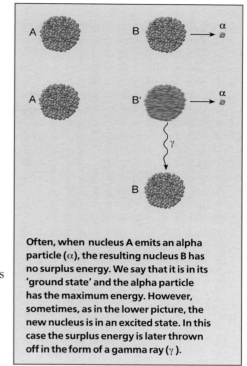

Often, when nucleus A emits an alpha particle (α), the resulting nucleus B has no surplus energy. We say that it is in its 'ground state' and the alpha particle has the maximum energy. However, sometimes, as in the lower picture, the new nucleus is in an excited state. In this case the surplus energy is later thrown off in the form of a gamma ray (γ).

Nevill Francis Mott, 1905–1996, driving, was one of the first to show how alpha particles, described by waves in alpha decay, appear as particle-like tracks in a cloud chamber (AIP Emilio Segrè Visual Archives, courtesy Sir Nevill Mott.)

minute compared to the energy required to make a cup of coffee. A quantum leap does not actually mean a big jump by everyday standards – the colloquial meaning of the phrase. However, it captures the idea that the nucleus undergoes a discontinuous change in energy – and also a discontinuous change in its structure. The atomic kind of quantum leap occurs on a very short timescale: sometimes a million millionth (10^{-12}) of a second after the one before, and sometimes a thousand times less (10^{-15} of a second).

Excited nuclei

Almost all nuclei have a series of excited states. The only exceptions are the few that are so fragile that they break up if they have too much energy. An example is the nucleus of deuterium, or heavy hydrogen, consisting of just a proton and a neutron bound together very weakly.

Each nucleus has its own unique pattern of possible excited energies. These 'fingerprints' are important in applications ranging from medicine to archaeology. Alpha decay is not the only process that leaves nuclei in excited states. Energy can be given directly to nuclei, for example, by bombarding them with fast particles from accelerators. The key point is that if a nucleus does find itself in a highly excited state then it will decay via a cascade of energy jumps, called transitions, until it reaches its lowest energy state.

When a nucleus makes a quantum leap between states, a gamma ray is emitted which carries off the energy difference between the states. These gamma ray

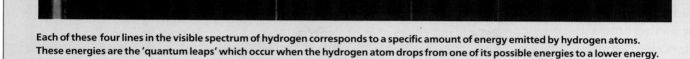

656.210 nm   486.074 nm   434.010 nm   410.12 nm

Each of these four lines in the visible spectrum of hydrogen corresponds to a specific amount of energy emitted by hydrogen atoms. These energies are the 'quantum leaps' which occur when the hydrogen atom drops from one of its possible energies to a lower energy.

energies can be measured very accurately with special detectors, allowing each nucleus to be identified from the pattern of gamma ray energies that it leaves behind, called its spectrum.

The same idea also applies to atoms and molecules where the energy of the emitted light corresponds to frequencies spanning the visible part of the electromagnetic spectrum. Green fireworks, for example, are that colour because they contain copper which has prominent green frequencies in its atomic spectrum, and when a pan of boiling, salted water spills onto a gas flame, the flame appears yellow due to the sodium in the salt.

This property is extraordinarily useful to astronomers who can use the spectra of atoms to determine what elements exist in distant stars and galaxies. They can also determine how common different elements are in the Universe.

Closer to home, some archaeologists study data from the spectra emitted from nuclei. These nuclear archaeologists can determine the precise location an artefact was made.

Wave–particle duality

Alpha particle radioactivity is a window into the strange behaviour of subatomic matter. What we have glimpsed through that window is so strange that there is still great debate about what it all really means today, a century after quantum physics was born. A natural reaction to hearing about the mysterious behaviour of the quantum world is to ask if it can really be like that. Yes, it can – there is now indisputable evidence that particles can sometimes behave like waves, and vice versa.

In alpha decay, both alpha particles and photons exhibit dual wave–particle behaviour. When a nucleus emits an alpha particle, the new nucleus which is formed may well be in an excited state in which case it will lose the excess energy by emitting packets of electromagnetic energy: gamma ray photons. Being electromagnetic radiation, gamma rays exhibit all the usual properties of waves; in particular they can create diffraction patterns. They also sometimes behave like particles: each photon is a package of energy and gives a single click in a Geiger counter, a pulse in a germanium crystal, or blackens a single grain on a photographic plate.

Alpha particles are helium ions, the very dense, electrically charged nuclei left after helium atoms have been stripped of their electrons. So far, we have concentrated on the particle nature of alpha particles. Each tiny flash of light on a zinc sulphide screen that eventually led Rutherford to his model of the nucleus marked the impact of an individual alpha particle. Similarly the tracks left in a cloud chamber are the result of particle behaviour. Yet alpha particles are as much waves as photons. Over the years there have been many experiments

**Henry Gwyn Jeffreys Moseley, 1887–1915.
(AIP Emilio Segrè Visual Archives,
W.F. Meggers collection.)**

4. Measuring nuclei
Determining the shape and size of minute objects

One of the most basic properties of nuclei is their size, and measuring something thousands of times smaller than the atoms of the measuring instruments poses problems. Yet by 1911 Rutherford knew something about the size of atomic nuclei. His discovery that they were some 10,000 times smaller than the atoms that contained them was due to the remarkable way that alpha particles scatter from metal foils. Nature had given us a wonderful and unexpected gift – a completely new way to learn about the deep secrets of matter.

Today we can go a long way towards determining the exact sizes and shapes of nuclei. What has been learned about nuclear sizes has become a cornerstone in our understanding of atoms. How this knowledge has been achieved is a story in which the oddness of the quantum realm plays an important part.

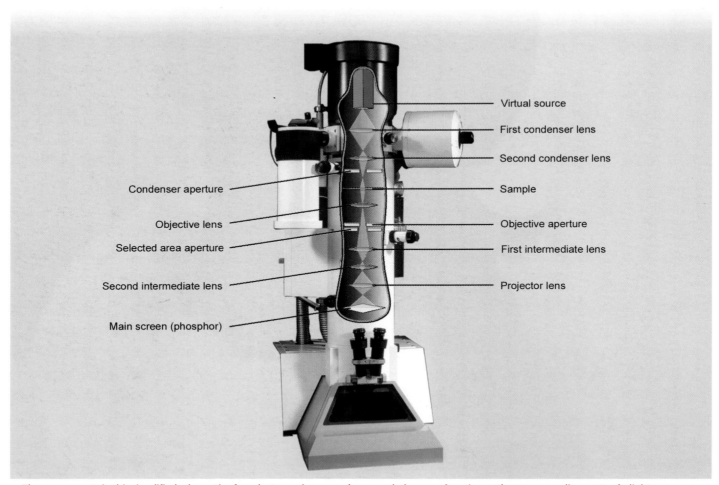

Condenser aperture

Objective lens

Selected area aperture

Second intermediate lens

Main screen (phosphor)

Virtual source

First condenser lens

Second condenser lens

Sample

Objective aperture

First intermediate lens

Projector lens

The components in this simplified schematic of an electron microscope have much the same function as the corresponding parts of a light microscope. The power of the electron microscope lies in the short wavelength of the electrons, allowing extremely fine details of the object to be resolved.

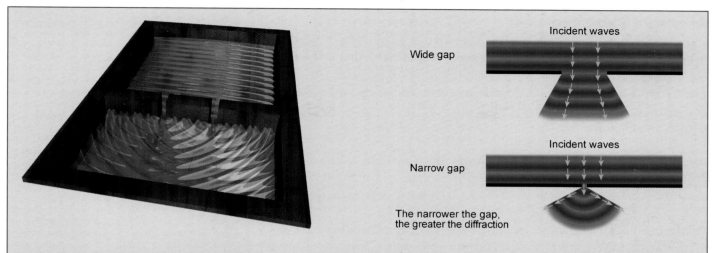

Waves 'diffract' (spread out) when they strike objects or pass through holes or slits in barriers. On the right is shown how the smaller the object or the hole relative to the wavelength, the stronger the diffraction, provided the object or hole is not much smaller than the wavelength. To study nuclei, we need much shorter wavelengths than can be achieved with an electron microscope.

Mega-microscopes

Previously we have described how it is not feasible to use ordinary microscopes in order to see objects smaller than a certain size. Instead, we depend on electron microscopes – which rely on the wave nature of electrons – to resolve objects as tiny as a virus, and even individual atoms.

The wavelength of electron waves can be made shorter if their energy is increased. So electrons need to be given enough energy to make their wavelength shorter than the details of the object under study. It is a big step from viewing viruses or individual atoms to seeing atomic nuclei, but it is one that has been taken. For achieving this step, a pioneer of the art, Robert Hofstadter, was awarded the Nobel prize in physics.

Hofstadter thought big. He had to, because the laws of physics demanded that large machines – what we might call 'mega-microscopes' – are needed to produce electrons with high enough energies to penetrate the secrets of the nucleus.

To understand how it is possible to gain an idea of the size of nuclei by observing matter waves, consider what happens when sea waves wash up against a hole in a breakwater. The smaller the hole, the more the waves are scattered, or spread out, on the other side. If you did not know the size of the hole, you could, in principle, work it out from seeing how the waves behave. If the distance between successive wave crests (the wavelength) is too large, then you do not get a pattern at all.

In a similar manner, light waves can produce a diffraction pattern when shone through various tiny holes. Again we can work out the size of the hole from the different patterns of light and dark rings.

The idea that particles like electrons really do act like waves was put to the test by Hofstadter and his colleagues in the 1950s when they asked for the large sums of money needed to build the first machines that accelerated electrons to high energies. To build such a machine was an act of faith, but their faith was rewarded. Today, thanks to experiments with particle accelerators, a great deal is known about a huge number of nuclei.

Robert Hofstadter, 1915–1990, pioneered the use of electrons for studying the size and density of nuclei and of the nucleons they are made of.
(Copyright the Nobel Foundation.)

Another quantum concept

You are probably becoming familiar with the idea that many common words do not mean quite the same when applied to atoms and nuclei as they do for everyday objects. 'Size' is one of these words. This is because, in the quantum world, objects such as nuclei simply cannot ever have sharp edges. Size is generally defined as the distance from one edge of an object to the other. If it does not have a definite edge, then its size is not clear-cut.

Nuclei are composed of protons and neutrons (nucleons) which obey the laws of quantum mechanics. According to these laws, we simply cannot say where a nucleon is in a nucleus; we can only say how likely it is to be found at a given point.

The laws of quantum mechanics are not just about uncertainty: they have a lot to say about how likely it is to find a nucleon in a given place. If you select some point in a nucleus, quantum mechanics determines the probability of a nucleon being at that point. This probability does not suddenly become zero at some particular distance from the nuclear centre. If it did, then that distance would be a sharp surface of the nucleus. Instead, nuclei have a diffuse (fuzzy) surface in which the probability of finding a nucleon drops off gradually.

Measuring fuzzy nuclei

How can this diffuse nuclear surface be measured? A proton carries positive electric charge, so the more likely that protons will be at a particular point in the nucleus, the more charge there will be at that point. The positive electric charge on the nucleus must therefore also be spread out in a way that has a fuzzy edge. Since electrons are electrically charged, a high speed electron moving through a nucleus is deflected by the positive charge in a way that depends on how that charge is distributed. This allows us to use a beam of electrons to measure a nucleus. It does not matter that the electron is unaware of the neutrons in the nucleus; it is sufficient that protons are evenly distributed throughout its volume.

Since electrons behave like waves, a beam of them striking a nucleus will produce a diffraction pattern. Although the electrons are detected as particles, the pattern is produced because greater numbers are found at certain angles and fewer at other angles. For water waves, both a barrier and a hole produce quite sharp diffraction patterns, but the patterns left by electrons scattering from nuclei are less distinct. The difference is that the hole and the barrier in the water have sharp edges and the water waves either pass through completely or not at all, but the fuzzy edge of the nucleus causes a gradual change in the quantity of deflected electrons. The details of the electron diffraction pattern reveal how the nuclear charge is spread out and we can determine just how fuzzy the nucleus is.

The size of atomic nuclei

To compare nuclear sizes we have to define what is meant by the word 'size' when applied to these diffuse objects. From electron scattering, we know nuclei have a rather constant density which falls off to zero at the surface in a way that is much the same for a wide range of nuclei. This allows a nuclear radius to be defined in spite of the fuzzy edge. The nuclear radius is the distance from the centre of the nucleus to a point where the density is half that at the centre.

Apart from the diffuse surface, all nuclei have about the same density. This can be deduced from the way that the volume of the nucleus depends on the number of nucleons it contains. Thus if nucleus A has twice the radius of nucleus B, then

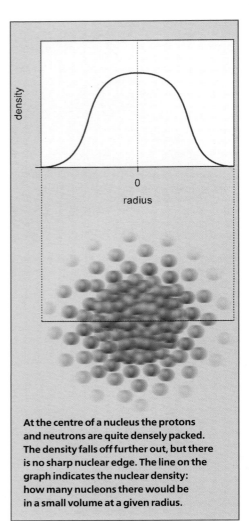

At the centre of a nucleus the protons and neutrons are quite densely packed. The density falls off further out, but there is no sharp nuclear edge. The line on the graph indicates the nuclear density: how many nucleons there would be in a small volume at a given radius.

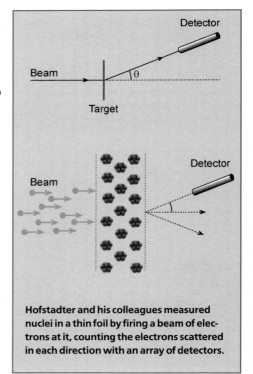

Hofstadter and his colleagues measured nuclei in a thin foil by firing a beam of electrons at it, counting the electrons scattered in each direction with an array of detectors.

nucleus A will have eight times (2 x 2 x 2) as many protons and neutrons as nucleus B. This is what is expected of everyday material that cannot be compressed. If a solid ball has twice the radius of another made of the same material, it would have eight times the mass. The volume, and hence mass, of a spherical object is proportional to the cube of its radius.

As nuclei obey this rule, it suggests all nuclei are made out of incompressible material of the same density. Atoms, however, are quite different.

It is remarkable that atoms and their nuclei behave in opposite ways: to a first approximation, atoms of different mass all have the same size and therefore different densities, whereas nuclei of different mass have the same density and therefore different sizes. There are exceptions in both cases: for instance alkali atoms like sodium are larger, and halogens like chlorine are smaller, but the difference is not dramatic. The fact that all atoms do not have the same density is crucial; if atoms of magnesium and of lead had identical density, then equal amounts of magnesium and lead material would also have the same density. We know this is not true – magnesium sinkers are not used for fishing, and lead wheels are inappropriate for fast cars!

For nuclei, however, the converse is true. A plutonium nucleus with about 240 protons and neutrons does have about twice the radius of a nickel nucleus containing an eighth as many protons and neutrons. This is because both nuclei have similar density, a density they share with almost all nuclei. The exception to this rule of uniform density is found in some remarkable cases that have caused a stir in recent years – the so-called 'halo nuclei', discussed later in this chapter.

The density of a nucleus is large compared with everyday objects. If a solid football were made of pure nuclear matter, it would weigh as much as Mt Everest.

Measuring nuclei with lasers

To measure nuclei using electrons, a target containing the atoms whose nuclei are to be measured has to be made, but not all nuclei are stable enough to sit in a target long enough for such experiments to be performed. There is a need to study such short-lived nuclei; they may well be different from stable nuclei, in which case a completely biased view of nuclear sizes would be obtained if only stable nuclei were studied. Other ways are therefore needed to measure nuclei.

There is a method that allows the measurement of nuclei that exist only fleetingly. It is based on the fact that, although the electrons in an atom are spread out through a vastly greater volume of space than the nucleus itself, there is nevertheless some probability that they can exist at the centre of the atom in the space occupied by the nucleus. This modifies the wavelengths of the light given out by the atom in a way which depends on the size of the nucleus. The effect is tiny, but lasers allow us to measure the wavelengths with incredible precision – enough precision, in fact, to turn the tiny effects of different nuclear sizes into a practical way of measuring them.

This method of measuring nuclei is known as the 'isotope shift method' because it is based on the way the wavelengths of light emitted by an atom varies from one isotope of the element to the next. Since different isotopes of the same element have the same number of electrons (and protons), their electrons will be distributed in almost exactly the same way. Where they do differ is in the number

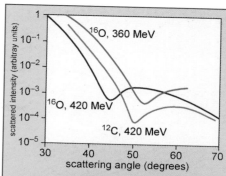

Just as the amplitudes of the waves at different angles allow the size of the hole to be measured in the figure on page 47, so the number of electrons found at each angle allow the size of the nucleus to be deduced. The energy of the incident electrons is in mega electron volts (MeV).

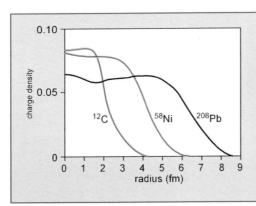

The lines show how the density of charge varies from the centre of the nucleus outwards. The nickel (Ni) nucleus of 58 nucleons is clearly larger than the carbon (C) nucleus of 12 nucleons, and the lead (Pb) nucleus of 208 nucleons is the largest.

of neutrons their nuclei contain, and this affects the size of the nuclei. Thus the light emitted by their electrons will differ very slightly due to the manner in which the electrons' probability distributions overlap the different-sized nuclei.

The difference in the wavelengths of light for different isotopes can be used to determine the variation in size of the nuclei. As long as the size of the nucleus of a stable isotope can be determined, for example by using the Hofstadter method of electron scattering, then the sizes of all the other isotopes of that element can be deduced.

Nuclear shapes

It is a natural assumption to imagine nuclei as tiny spheres. In the 1930s, even before lasers were invented, crude isotope shift measurements were used on different isotopes of the fairly rare element, samarium. They showed that nuclei of the heavier samarium isotopes appeared larger than expected. We now know this is because they are not spherical. Today it is known that quite a large proportion of nuclei have non-spherical shapes. Such nuclei are said to be deformed.

Most of our evidence about nuclear shapes arises from the fact that non-spherical nuclei can be made to spin. Another rather odd quantum concept is that if a quantum object, such as a nucleus, is perfectly spherical, then there is nothing to distinguish its orientation in space. It is meaningless to state whether or not a spherical nucleus is spinning, since we could never tell.

A = 8	A = 27	A = 64	A = 125
$R = 2R_0$	$R = 3R_0$	$R = 4R_0$	$R = 5R_0$

How does the nuclear radius depends on the number of nucleons? Most nuclei have the same density, so their volume is proportional to the number of nucleons. As a result, the cube of the radius is also proportional to the number of nucleons, as we see here.

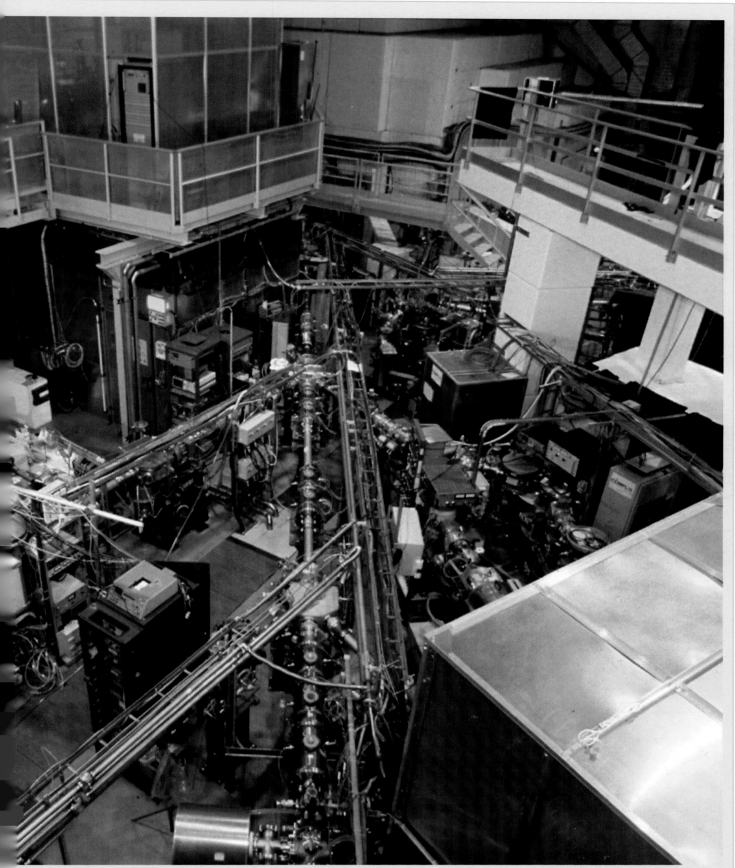

To study the properties of a nucleus that lives for a thousandth of a second, it is necessary to do 'on line' experiments in which nuclei are created and studied in the same apparatus. This is ISOLDE (Isotope Separator On Line) at CERN in which such experiments were pioneered. In this picture you can see the beam lines down which the radioactive ions are steered in a vacuum towards the experimental stations. (CERN)

The more deformed a nucleus is, the more easily it can be made to rotate. This rotation is another way in which the nucleus can store energy – energy it can lose by emitting gamma rays and hence slowing down. The more deformed it is, the more quickly a spinning nucleus slows down by radiating gamma rays. By measuring how easily nuclei are set spinning when other nuclei are fired at them, or by seeing just how quickly spinning nuclei slow down by emitting gamma rays, the amount they deviate from the spherical can be determined.

Most nuclei have one of two basic shapes. They are either spherical or prolate (shaped like an American football). An important exception is the nuclei of carbon, which are oblate (like a squashed sphere). Both prolate and oblate nuclei are nearly always axially symmetric.

Quantum mechanics has another trick up its sleeve concerning deformed nuclei. A deformed nucleus can no more lie with its long axis fixed pointing in a given direction than a proton or neutron can sit still at some fixed point within the nucleus. Both are a result of the Heisenberg Uncertainty Principle. Thus, the long axis of a prolate nucleus has equal probability of pointing in every direction. In other words it is pointing in all directions at once!

If an isotope shift experiment is performed on an atom containing a prolate nucleus, the electrons in the atom see a fuzzier nucleus than usual. The tip of the long axis of the nucleus traces out a larger volume of space than any point on the surface of a spherical nucleus with the same mass because it lies farther out. As a result, the surface is even more uncertain than for spherical nuclei, causing a shift in the spectra.

Modern experiments reveal more about the shape of deformed nuclei than just whether they are prolate or oblate. Amongst the many nuclei which are strongly deformed are the isotopes of uranium and plutonium, a fact important for the process of nuclear fission.

Super-deformed nuclei

Most prolate-shaped nuclei are only slightly deformed. However, in the mid 1980s, physicists at the Daresbury Laboratory in the UK discovered a remarkable phenomenon: super-deformation. Sometimes nuclei are twice as long as they are wide.

When nuclei rotate they have a tendency to stretch. This is due to the same force you feel when sitting in a fast car as it turns a tight corner. Nuclei are generally too tough to stretch very much, but at Daresbury it was found that if certain nuclei are made to rotate really fast, it is as if something suddenly gives and the stretching force takes the nucleus into an entirely new shape. The nucleus becomes super-deformed. As they slow down, by emitting lots of gamma rays the nuclei revert back to the normal deformed or even spherical shapes.

Halo nuclei break the rules

Although most nuclei have the same density, there are exceptions. One such nucleus, the deuteron, has been studied for a long time. The deuteron is the nucleus of heavy hydrogen, consisting of just one proton and one neutron that are very weakly bound together. As a result of this weak binding, the proton and neutron have a significant probability of lying very far apart. The probability distribution of the deuteron extends out in space to such an extent it has the same size as that of a

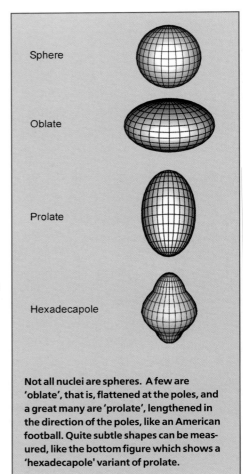

Not all nuclei are spheres. A few are 'oblate', that is, flattened at the poles, and a great many are 'prolate', lengthened in the direction of the poles, like an American football. Quite subtle shapes can be measured, like the bottom figure which shows a 'hexadecapole' variant of prolate.

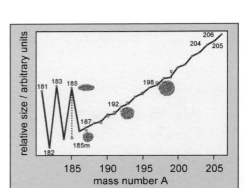

Mercury (Hg) isotopes with decreasing masses steadily decrease in radius. But at mass 186, 184 and 182, the removal of another neutron produces a jump in size. The anomalously large size of these nuclei is evidence that they are not spherical. These measurements were a triumph for ISOLDE. The isotope ^{185}Hg has an excited state which, unlike the ground state, has the normal size.

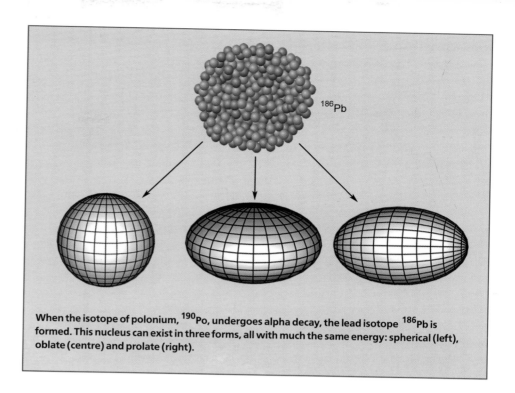

When the isotope of polonium, ^{190}Po, undergoes alpha decay, the lead isotope ^{186}Pb is formed. This nucleus can exist in three forms, all with much the same energy: spherical (left), oblate (centre) and prolate (right).

Ben Mottelson, 1926–, American-born physicist living in Denmark. He was awarded the Nobel Prize in 1975 with James Rainwater, 1917–1986, and Aage Bohr, 1922 –, the son of Niels Bohr, for work on vibrations and rotations of nuclei. (Copyright the Nobel Foundation.)

very much heavier nucleus. This is where we catch another glimpse of how strange the quantum world is: the proton and neutron spend part of the time outside the range of the force that holds them together, yet they still remain bound together unless the deuteron is hit by something like a powerful gamma ray.

Adding a second neutron to a deuteron makes a nucleus of tritium, an even heavier isotope of hydrogen. This extra neutron draws the original proton and neutron closer together so the tritium nucleus (a triton) is actually smaller than the deuteron. Adding a second proton to a triton continues this process, making a tightly bound helium nucleus – an alpha particle. The two protons and two neutrons in the alpha particle are so close together that this nucleus has much the same density as most heavier nuclei. Subsequently, adding protons or neutrons does not affect the nuclear density.

In recent years a remarkable new class of nuclei has been discovered, the halo nuclei. The most famous is a lithium isotope with no less than eight neutrons accompanying its three protons. This is ^{11}Li. The last two neutrons in this nucleus are very weakly bound, and are spread out in a volume which extends far outside the range of the other nine nucleons. As a result, this nucleus is as large as a nucleus of lead with over 200 protons and neutrons.

The discovery of halo nuclei in the mid-1980s electrified the community of nuclear physicists. These nuclei present a rich variety of phenomena to study, and stretch our understanding of nuclei theory to the limits. As halo nuclei have only a fleeting existence, they need special facilities to study them. Halo nuclei are now studied in many laboratories around the world.

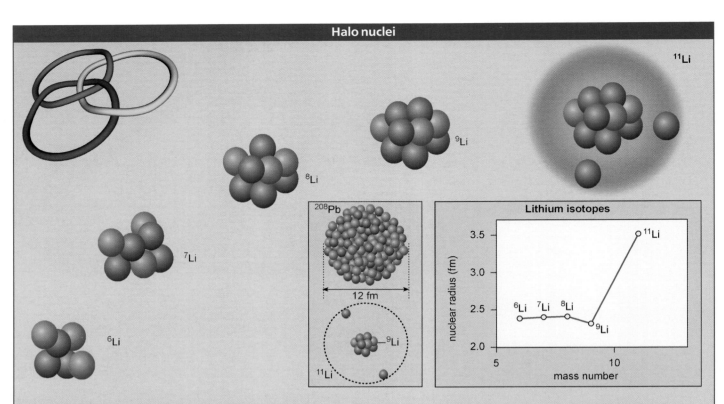

The isotope of lithium with three protons and six neutrons has just the size that would be expected for a nucleus with nine nucleons. There is no stable lithium nucleus with seven neutrons, but there is a lithium nucleus with eight neutrons ... and it stretches out as far as a lead nucleus. Its surface is very tenuous, containing just two neutrons in a large volume. Several of these 'halo nuclei' have now been found, and they present a huge challenge to experimenters and theorists struggling to understand their many strange properties. The Borromean rings at top left are such that if one is cut, all three fall apart. This is also true of the ^{9}Li core and two outer neutrons that make up the halo nucleus ^{11}Li. ^{10}Li does not exist, nor does a nucleus containing just two neutrons.

5. Strange nuclear material
The contents of a nucleus

So far, an atomic nucleus has been described as a collection of Z protons and N neutrons. The protons each carry one unit of positive charge, whereas the neutrons are uncharged and are very slightly heavier than protons. A nucleus has been assumed to contain only protons and neutrons. However, enough quantum mechanical surprises have been introduced to suggest that, when it comes to nuclei, nothing is as simple as it seems.

When nuclei fall apart

One way of finding the constituents of nuclei is to observe what is emitted from them. In radioactivity, some nuclei spontaneously eject particles and change into different kinds of nuclei without any prompting. Stable nuclei, however, need inducement before they will eject particles. In fact, any nucleus can be broken apart by hitting it hard enough with another nucleus. This can be done in the laboratory by using high energy particles from accelerators, but it also happens naturally in stars.

Nuclear reactions were important even before stars existed; most of the nuclei in the Universe were created a few minutes after the Big Bang. At that time, the Universe was an extremely hot and crowded place, with nuclei being formed and, almost as fast, broken apart. Almost immediately, the Universe expanded and cooled sufficiently for nuclei not to be broken apart as soon as they were created. These surviving nuclei formed the primordial material which eventually condensed into stars.

Radioactivity and the constitution of nuclei

The idea that most of the mass of atoms resides in compact nuclei was still years in the future when Rutherford first noticed that radioactivity involved two kinds of particles. It soon became clear that beta rays were J.J. Thomson's recently discovered electrons, since their paths were bent by magnetic fields in exactly the same way. They also penetrated matter much more readily than alpha particles. It took a little longer to identify alpha particles as helium nuclei.

One possible conclusion to be drawn from radioactivity is that atomic nuclei are made of alpha particles and electrons, but this would not work for hydrogen since the hydrogen nucleus has much less mass than an alpha particle. To solve this problem, a very old idea was revived: that all atoms are composed of multiples of hydrogen. This idea originated with William Prout in 1815, and is known as Prout's hypothesis. It was applied a hundred years later to nuclei rather than atoms; all nuclei were predicted to be composed of various numbers of hydrogen nuclei (protons).

This theory cannot be quite right as, for example, the nucleus of a nitrogen atom has the mass of fourteen protons, but the charge of just seven protons. The textbooks of the 1920s had a simple answer to this: the nitrogen nucleus has fourteen

Do nuclei have alpha particles within them? By far the most common isotope of all the even Z elements from carbon to sulphur have the same number of protons and neutrons as three, four, five etc. alpha particles. It is not true for argon for which ^{36}Ar has an abundance of less than 1%. These facts are best explained in terms of the the energy valley (discussed in Chapter 6) and the tendency of protons and neutrons to pair up. Before the discovery of the neutron, it was natural to suppose that nuclei are composed of protons, electrons and alpha particles.

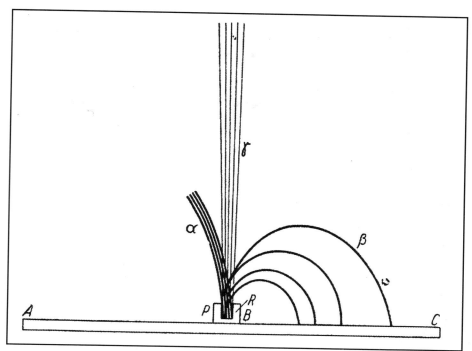

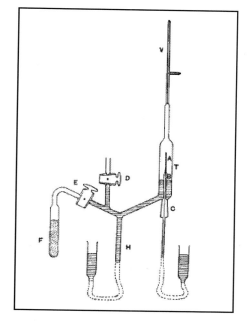

The apparatus used by Rutherford and Royds in 1908 to prove that alpha particles were helium ions (in modern terms: helium nuclei). Alpha particles penetrated through a thin window into a vacuum chamber. When an electric discharge was passed through this chamber the light given out was analysed by a spectroscope, which confirmed the presence of helium.

protons and also seven electrons to balance out the charge of half the protons. After all, there must be electrons inside atomic nuclei since they come out of them in beta decay.

In this proton–electron model, alpha particles would consist of four protons and two electrons. Many nuclei decay by ejecting this particular structure from the nucleus as a complete particle, so there must be something special about this configuration. Another feature also suggests the helium nucleus is somehow special: the most common isotopes of many light elements are apparently multiples of the alpha particle. By far the most common isotope of carbon, for example, is ^{12}C which has the same charge and mass as three alpha particles. ^{13}C is very uncommon and ^{14}C is unstable. From this, the alpha particle appears to be special in some way, and indeed some nuclei do contain alpha particles but not in a straightforward way, as we see at the end of this chapter.

Do nuclei contain electrons?

Like alpha particles, electrons appear to be ejected from nuclei. Unlike alpha particles, there is a very straightforward answer to the question of whether they exist in nuclei. They do not, and they cannot.

^{12}C	^{16}O	^{20}Ne	^{24}Mg	^{28}Si	^{32}S	^{36}Ar	^{40}Ca
3α	4α	5α	6α	7α	8α	9α	10α
98.9%	99.8%	90.5%	79.0%	92.2%	95.0%	<1%	96.9%

Firstly, the absence of electrons within nuclei can be understood from Heisenberg's Uncertainty Principle. This principle is commonly, and wrongly, taken to mean that everything on the microscopic level is uncertain. In fact, it implies that the more we try to localize a quantum particle such as an electron – that is to confine it to a smaller and smaller region of space – the larger the fluctuations in its motion due to the uncertainty in its momentum. The result is that for an electron to be confined to the tiny nucleus, its motion would be too violent for it to stay there for long. Protons on the other hand, which are nearly two thousand times heavier than electrons, can easily be confined to the volume of a nucleus since they move about more slowly and the uncertainty in their motion is much less.

Another argument against atomic nuclei having any electrons in them comes from the light emitted when an electric discharge is passed through nitrogen gas. The fine details in the pattern of spectral lines indicate clearly, according to certain quantum rules, that the nitrogen nucleus has an even number of particles, but if this nucleus is made of protons and electrons it would have to be an odd number: fourteen protons plus seven electrons.

In 1932, Chadwick's discovery of the neutron, which is a neutral particle with about the same mass as a proton, solved this problem. Werner Heisenberg then suggested, correctly, that atomic nuclei consist purely of protons and neutrons. Chadwick's discovery was to mark the beginning of modern nuclear physics. Now the nitrogen nucleus would have seven protons and seven neutrons: an even number of particles in total.

The question now arose that if nuclei consist of just protons and neutrons, where do the beta decay electrons come from? These electrons also posed another problem, one of the greatest problems of physics in the twentieth century, and a problem which drove physicists to desperate remedies.

Missing particles

Beta decay electrons appeared to break the law of conservation of energy. The nucleus emitting the electron has a definite energy, and the nucleus produced in the decay has a definite energy, but the electrons emerge with a range of energies always less than the difference in energy between the initial and final nuclei. The missing amount of energy suggested energy is not always conserved in beta decay, but as the law of conservation of energy is so sacred in physics, few physicists dared to speculate that there was an exception to the rule. Despite this, Niels Bohr did just that – and turned out to be wrong.

Another proposal, made by Wolfgang Pauli in 1930, was that a second, virtually undetectable particle was emitted at the same time as the electron. This new particle, later called a neutrino (meaning 'little neutral one'), shared energy with the electron in a random way such that the total energy of the two particles was equal to the difference in energy between the parent and daughter nuclei. In this manner, energy is conserved after all.

Pauli's idea was radical, but it became a key part of a theory of beta decay proposed by the brilliant Italian physicist, Enrico Fermi, in 1934. Over the years, Fermi's theory has been greatly extended and incorporated into more modern models, but still successfully accounts for a wide range of phenomena connected with beta decay. According to this theory, in a nucleus that has more than the ideal number of neutrons, a neutron turns into a proton, creating both an electron and a neutrino.

James Chadwick, 1891–1974. His discovery of the neutron in 1932 was a turning point in nuclear physics. Chadwick was well aware that the neutron had been predicted by Rutherford in 1920. (Copyright the Nobel Foundation.)

The first evidence for the positron, the anti-particle of the electron. The cloud chamber track bends the wrong way in a magnetic field for it to be caused by an electron, and it is not a proton track. Anderson's unprecedented discovery was highly controversial until it was confirmed a year later. (Courtesy C.D. Anderson.)

Neutrinos are exceptionally hard to detect. More than 150 million neutrinos pass through every square centimetre of your body every second, and even a body the size of the Earth will very rarely stop a neutrino. It is therefore not surprising that it was 1956 before Reines and Cowan could announce neutrinos had been positively detected. This was long after Pauli's speculation (1930) and Fermi's theory (1934), but proved that they had been correct. Today, neutrinos play a vital role in astronomy and cosmology as well as in nuclear physics.

Antimatter

Fermi's theory embodies one of the most revolutionary ideas of twentieth century physics: the number of fundamental particles in the world is not fixed. Electrons are not emitted from nuclei in beta decay, but are created, along with neutrinos. The idea that particles can be created was one consequence of combining quantum mechanics with Einstein's special theory of relativity, a feat achieved by Paul Dirac.

Combining relativity and quantum mechanics had another consequence: the existence of anti-particles. Dirac's theory predicted that for every kind of particle there is a corresponding anti-particle having the same mass. If the particle has an electric charge, the anti-particle has the opposite charge, but even neutral particles like neutrons have anti-particles. The anti-particle for the electron, called the positron, was discovered in 1932 by Carl Anderson.

Enrico Fermi, 1901–1954, was unique in the 20th century for his eminence in both theoretical and experimental physics. (AIP Emilio Segrè Visual Archives.)

When a particle meets its anti-particle, they annihilate each other releasing pure energy that is carried off as radiation. The energy released is that predicted by Einstein's famous equation, $E = mc^2$. When an electron and a positron annihilate each other, for example, the energy released, E, is equal to their combined mass, m, multiplied by the square of the speed of light. The process of annihilation is reversible and when a high energy photon, maybe from cosmic rays, interacts with a nucleus it quite often disappears, with its energy reappearing as two particles, an electron–positron pair.

Beta decay as it was originally discovered takes place when nuclei with too many neutrons correct the imbalance by transmuting a neutron into a proton, an electron and an anti-neutrino. Nuclei with the opposite imbalance, too many protons, decay in the opposite fashion: a proton is converted into a neutron, a positron and a neutrino. This kind of beta radioactivity was first discovered by Irène Joliot-Curie (Marie Curie's daughter) and Frederic Joliot in 1934 and is now of immense importance in modern medicine, particularly in positron emission tomography (known as PET for short). PET is a much used diagnostic tool, especially important for diagnosing brain disease.

Chadwick's discovery of the neutron provided the foundations for a very successful picture of nuclear structure which has been the cornerstone of nuclear physics for many years. In this model, nuclei consist of protons and neutrons, and the emission of electrons and positrons does not mean that these particles are contained within them.

Carl D. Anderson, 1905–1991, discoverer of the positron. (AIP Emilio Segrè Visual Archives.)

Owen Chamberlain, 1920–, together with Emilio Segrè, discovered the antiproton, the negatively charged anti-particle of the proton, in 1955. (Copyright the Nobel Foundation.)

Paul Adrien Maurice Dirac, 1902–1984, left, and Richard P. Feynman, 1918–1988. Dirac was the first to successfully combine quantum theory and relativity; in doing so, he predicted the existence of positrons. Feynman developed a formalism for doing the very complex calculations involving electrons, positrons and light. (AIP Emilio Segrè Visual Archives.)

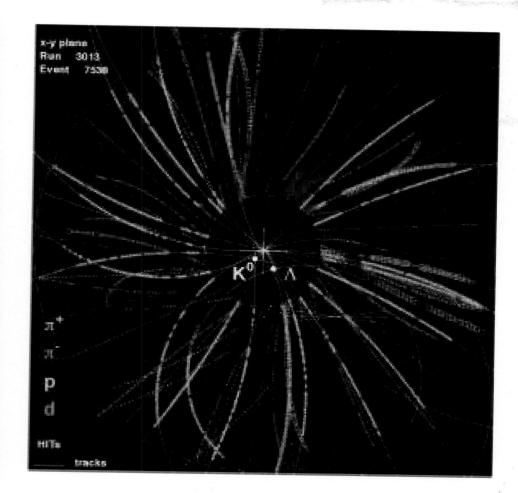

x-y plane
Run 3013
Event 7538

K^0 Λ

π^+
π^-
p
d

Hits

tracks

When two nickel nuclei collide at 89% of the speed of light, many particles are released. This picture, taken using a 'drift chamber' at GSI in Darmstadt, Germany, allows them to be identified. Two of these particles are exotic: a K-zero (K^0) and a lambda (Λ). They did not exist in the incoming nuclei, and very quickly decay into more familiar particles. Such experiments yield vital clues concerning neutron stars and supernova explosions. (Courtesy GSI.)

When nuclei collide

In 1919, Rutherford discovered that when alpha particles from a radioactive source travel through air, a proton is sometimes produced. The alpha particles had knocked protons out of the nuclei of nitrogen atoms in the air – the first time a nuclear reaction had been observed.

Nuclear reaction experiments are no longer performed with alpha particles from radioactive sources. Large accelerators produce beams of a wide variety of particles, ranging from electrons and protons to uranium nuclei. The energy of these projectiles is very much higher than alpha particles emitted in radioactive decay and can be precisely controlled. In addition, the number of particles per second is vastly greater than a radioactive source could possibly produce. Beams of high energy particles can be directed onto targets containing any chosen nuclei, where they collide and nuclear reactions take place.

When two nuclei collide at high energies, many different processes occur, leading to the production of a variety of final products. Sometimes the resulting nuclei have a great deal of internal energy as a result of the collision. By choosing suitable beams and targets, it is possible to select what new types of nuclei will be produced in order for these to be studied. Some of these products have found a vital role in modern medicine, and many have quite different properties to the nuclei of atoms found naturally on Earth. However, as long as the energy of the incident particle is not too high, all the transmutations are consistent with the proton–neutron model: the total numbers of protons and neutrons in the products of the collision are the same as those in the original. If the energy of the collision is higher, other particles can be produced.

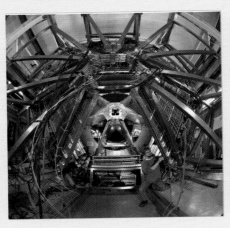

Taking nuclei apart

One particular kind of reaction is known as photo-disintegration which, as the name suggests, involves the breaking up of nuclei by light. Just as the ultraviolet radiation in sunlight breaks up dye molecules and, in so doing, fades clothes and fine tapestries, so high energy photons can break up nuclei. These are not photons of visible light, but high energy gamma rays, the same radiation that nuclei emit when they lose energy by jumping down from one quantum energy level to another.

Just as there are machines for making high energy particles of matter, so there are machines that produce high energy gamma ray photons. However, when Chadwick and Goldhaber carried out photo-disintegration for the first time, breaking up a deuteron into a proton and a neutron, they used gamma rays from

Top: Aerial view of the Relativistic Heavy Ion Collider (RHIC) at Brookhaven National Laboratory in the United States. Heavy nuclei collide head on at close to the speed of light to produce a quark–gluon plasma. (Brookhaven National Laboratory)

Below: The outer pictures show part of the HADES spectrometer being built at GSI near Darmstadt in Germany to study the electron –positron pairs produced when heavy nuclei collide at high energies. This will help us to understand nuclear matter at high density. The central picture is a view inside the linear accelerator which gives the nuclei at GSI their first boost towards high energy. (Courtesy A. Zschau, GSI.)

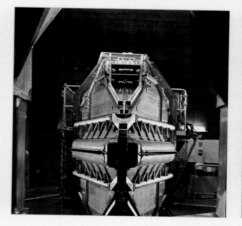
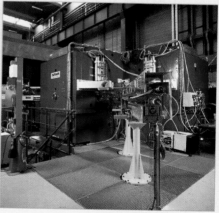

radioactivity. Today, with a machine like the one at Mainz, in Germany, it is possible to knock a proton or a neutron out of any nucleus with photons or electrons. It is easy to imagine doing this repeatedly to a nucleus, one nucleon at a time, until ending up with a deuteron, and then finally just a proton or neutron. If it is possible to take a nucleus apart one nucleon at a time in this way it suggests that nuclei must contain only protons and neutrons.

Quantum mechanics, being quantum mechanics, implies matters are not so simple.

Top: The Mainz Microtron (MAMI), an accelerator at Mainz in Germany, produces high energy electrons for probing fine details of nuclear structure. (Johannes Gutenberg-Universität Mainz)

Below: Three images from GANIL, a complex of accelerators at Caen in France devoted to experiments with beams of heavy nuclei. The central picture shows the main cyclotron (CSS), on the left is a resonator, and on the right is the CIME cyclotron which accelerates exotic nuclei produced using the CSS. (Copyright M. Desaunay, J.M. Enguerrand/GANIL.)

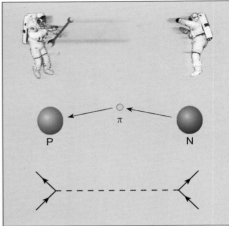

If two astronauts in space were to throw spanners to each other, it would tend to force them apart. Nucleons are not like this. The reason why particle exchange causes attraction lies in the depths of the mathematics. The bottom section, with the dashed line indicating the exchanged meson, shows the diagram invented by Feynman that symbolizes the exchange process.

Hideki Yukawa, 1907–1981, put forward a theory for the force that holds protons and neutrons together based on the exchange of a new kind of particle, the meson. The first meson to be discovered is now known as a pion. (AIP Emilio Segrè Visual Archives, W.F. Meggers Collection.)

Something else inside nuclei

Particles can be created and destroyed, like the electrons of beta decay which are created in the decay process, or the electron–positron pairs created from high energy photons. In 1935 the Japanese physicist Hideki Yukawa proposed a radical, and Nobel prize winning, theory for the attractive force between nucleons. This is the force that binds a proton and a neutron into a deuteron and holds together all nuclei. Yukawa proposed that the origin of this force is an entirely new kind of particle. It is created at one of a pair of neighbouring nucleons and jumps across at near the speed of light to be absorbed by the other nucleon. It may seem surprising that this gives rise to an attractive force, but such an attraction was a clear prediction of Yukawa's calculations. There are also other examples. The electric force between charged particles (repulsive for like charges, attractive for unlike charges) is also due to exchange of particles, but in this case the particles are photons.

Yukawa's particle was eventually discovered more than a decade later, and is now known as the pion; it is the lightest of a class of particles known collectively as mesons. Originally, mesons were so-called because they were intermediate in mass between electrons and protons, the pion being about 273 times as heavy as an electron, whereas a proton has about 1835 times the mass of an electron. Over the years many other mesons have been found, including some so heavy they have the mass of a large nucleus.

According to the modern versions of Yukawa's theory, the nucleons in the nucleus are held together by mesons coming into a very brief existence during which they jump from one particle to another before vanishing again. The energy to create the meson is a loan from a quantum bank and has to be very quickly repaid. So, at any given time nuclei must contain some mesons as well as protons and neutrons.

Nature allows energy to be borrowed briefly to allow an exchange meson to be created out of nothing. The law of energy conservation can be challenged only for a VERY brief period; the bigger the energy loan, the more quickly it must be paid back.

Mesons cannot explain everything about the force between a pair of nucleons, especially when they are very close together. A fuller understanding involves the fact that protons and neutrons, like all the mesons, are themselves composite objects. They consist of even more fundamental building blocks, called 'quarks', which are held together within the nucleons and mesons by the exchange of particles known as gluons. Modern nuclear research has opened up the whole question of how nuclei can provide a new window into Nature at the quark–gluon level.

No simple answer

When very high energy projectiles collide with other nuclei, they sometimes apparently knock pions out. However, this is not strictly evidence that there are pions inside nuclei since pions can also be created out of the energy of the incident projectile. As has been seen previously, the fact that particle 'X' comes out of nuclei is not proof that nuclei have particle 'X' inside them. There are, however, some rather difficult experiments involving beams of electrons and photons which can only be understood if it is assumed that there are pions within nuclei.

It is quite typical of the quantum world that the answer you get to any question depends on how you ask it. If you ask what is in the nucleus, and ask the question with low energy particles, then the answer is: just protons and neutrons; in such a low energy experiment if a nucleus is taken apart nucleon by nucleon, it can never end up with just a pion. Asking the question with high energy particles, the answer will be different and will include a whole variety of other particles.

Mesons do not end the list of what might be in nuclei. Atomic nuclei, like atoms, have excited states and if a nucleus is in one of its excited states, it will quickly lose its excess energy by emitting gamma rays and jumping down to the state with the least energy, the ground state. In the 1950s it was found that protons and neutrons can also be excited to higher energy states whereby their properties are changed and they become new particles.

The most common state that an excited nucleon can be in is known as a 'delta resonance', or a delta. Outside nuclei, deltas live for a very short time (about the time it takes for light to cross from one side of a proton to the other), but inside a nucleus, a proton can take a very short-term loan of energy from the same mysterious quantum bank that provides energy for mesons. Protons and neutrons use this energy to fund a brief transformation into delta particles. The result is that the proton within a deuteron is actually a delta for around one percent of the time.

It is possible to knock deltas out of a deuteron, but only by hitting the deuteron with enough energy to turn the nucleon into a delta. But only pions come out to be seen in detectors. This is not evidence that there are deltas ready-formed inside deuterons. The evidence for deltas is either indirect or due to the fact that certain nuclear theories depend for their success on deltas, and other types of particles, being formed from excited states of the nucleon, and existing fleetingly in nuclei.

Nuclear matter at high energies

We have seen that ordinary nuclei contain not only protons and neutrons but also mesons and other particles which exist very briefly before disappearing altogether when the nuclei are dismantled with low energy reactions. Examining reactions at high energies is also important because there are locations in the Universe where nuclei, or the matter of which nuclei are composed, exist in extraordinary states. Neutron stars, for example, are composed of nuclear matter, but not quite the

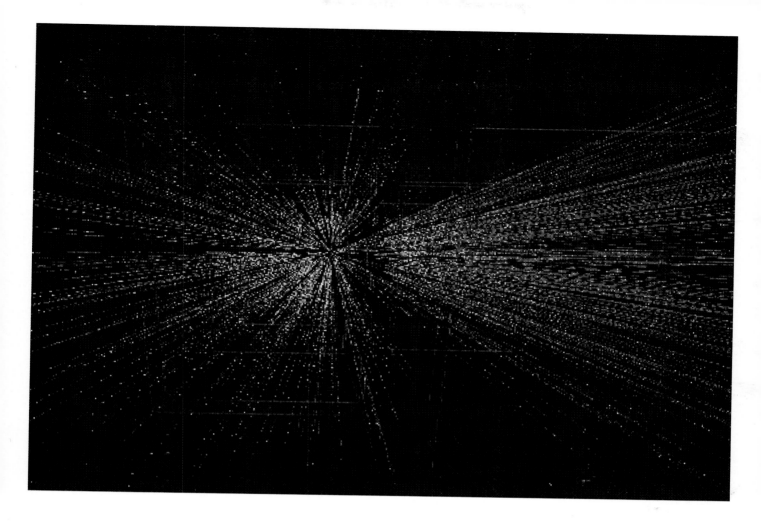

When two lead nuclei collide at the highest possible energy, a quark–gluon plasma is fleetingly produced. It then disappears in this spectacular shower of particles as registered in a track chamber. The beam direction is towards the observer. The Universe consisted of a quark–gluon plasma a few microseconds after the Big Bang. (CERN)

Tracks left by high energy particles in a bubble chamber. Some particles are bent much more easily by a magnetic field than others, just as in the diagram from Marie Curie's thesis on page 57. A little to the left of the centre, a 'V' shows a particle–antiparticle pair created from a neutral particle that has left no track. (CERN)

same type of nuclear matter as the ingredients of ordinary nuclei in their ground states. The stuff of neutron stars cannot be made on Earth, but clues about it can be gained from the study of what happens when very high energy nuclei collide. Not surprisingly, many new particles are created.

When two complex nuclei collide at high energy, many different kinds of mesons besides pions emerge, and these give clues as to the behaviour of nuclear matter at high energy and pressure. This information is needed by physicists who are trying to understand neutron stars and the processes which take place in supernovae, the explosions of stars which are the source of many of the elements found on Earth.

There does not appear to be a limit to the variety of particles that can be made to come out of nuclei if enough energy is pumped in. The vast array of particles which emerge in the highest energy collisions has recently provided evidence that in very hot nuclei the individual protons and neutrons melt away leaving a soup of quarks and gluons, the so-called 'quark–gluon plasma'. It is important to understand this state of matter since conditions in the very early Universe were once fleetingly extreme enough for it to exist.

Quantum tunnelling

The suggestion that whatever comes out of nuclei must be inside them has already been shown not to be true for electrons. The question of whether this is true for alpha particles reveals, yet again, that the quantum world makes the answer (both yes and no) extremely intriguing. Alpha particles are tightly bound nuclei with their two protons and two neutrons fitting together very snugly. It is for this reason that they often have an independent existence, of a sort, within many nuclei.

The nucleus of ^{20}Ne, the isotope of neon composed of ten protons and ten neutrons, has a structure similar to five alpha particles. This is predicted even by models which assume the twenty nucleons move freely throughout the nucleus. Although the nucleons within certain nuclei do tend to group themselves into alpha-like clusters, the clusters do not consist of four particular nucleons, or always the same four nucleons. In quantum mechanics, every proton in the nucleus is partly in every alpha-like cluster and the same is true for every neutron. The fact that all protons are identical in the special quantum mechanical sense means that every proton is equally part of every cluster. Thus the simple intuitive picture of alpha particles bouncing around in nuclei, each with its own unique set of nucleons, is not quite accurate.

A heavy nucleus like the uranium isotope ^{238}U, which decays by emitting alpha particles, has many more neutrons (146) than protons (92). It therefore cannot be thought of as a collection of alpha particles. Nevertheless, alpha particle-like structures do occur in the nucleus and sooner or later these appear outside the nucleus as alpha particles, and can be picked up in detectors.

The greater the probability that the alpha particles are emitted, the shorter the half-life of the nucleus. The half-life of a nucleus such as ^{238}U is determined by two factors. The first is the likelihood that the protons and neutrons arrange themselves into structures that look like a nucleus of the thorium isotope ^{234}Th, plus a single alpha particle. The second factor is how rapidly the alpha particle can tunnel out of the nucleus.

In 1928 George Gamow calculated the tunnelling probability of an alpha particle. It is the different rates of alpha tunnelling that determines the huge difference in half-lives: 4.5 billion years for ^{238}U and a third of a microsecond for the polonium isotope ^{212}Po. The essential point of quantum mechanical tunnelling is that quantum theory makes it possible for particles to tunnel through a barrier which it does not have enough energy to jump over. Such a barrier would be absolutely impenetrable according to pre-quantum physics.

The ease with which a particle gets through this barrier depends extremely sensitively on the amount of energy possessed by the particle. Just how sensitively can be seen from the huge difference in the half-lives of the uranium and polonium isotopes. The vast ratio of half-lives corresponds to a factor of just two in energy: the alpha particle emitted by ^{212}Po has about twice the energy of the alpha particle emitted by ^{238}U.

The importance of tunnelling in nuclear physics extends far beyond understanding alpha decay. The fusion reactions by which stars get their energy and by which elements are built up from hydrogen, depend on tunnelling inwards through otherwise impenetrable barriers. The sensitivity of the rate of tunnelling to the energy of the particles is a crucial ingredient in governing the rate at which stars burn and evolve. Tunnelling makes this possible and governs not only the energy from the Sun, but is a key to producing the very elements from which we are made.

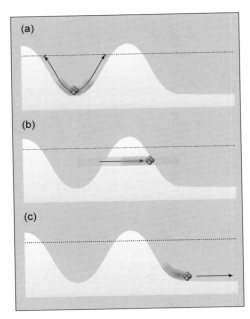

(a) shows a particle (imagine a marble) rolling back and forth simply does not have enough energy to get out of the well. However, quantum mechanics allows something amazing to happen: a particle with too little energy to get over the top can occasionally tunnel through the well (b) in a way that has no parallel in the world of human-sized objects. After tunnelling through, the particle accelerates 'down the hill' just as everyday experience would suggest (c).

The Thomas Jefferson National Accelerator Facility in Virginia, United States, represents a new step in electron accelerators. It is already producing new information about nuclear structure at the quark level. (Courtesy the Thomas Jefferson National Accelerator Facility.)

6. The nuclear landscape
The variety and abundance of nuclei

It is obvious, just by looking around, that some elements are much more abundant on Earth than others. There is far more iron than gold, for example – if it were the other way round, both the ship-building and jewellery businesses would be very different. In addition, it is not just on Earth that some elements are rare. Astronomers have found there is far more iron than gold throughout the Universe, although even iron is relatively rare since it, along with all other elements apart from hydrogen and helium, comprise just 2% of all the visible matter in the entire observable Universe. The other 98% is about three-quarters hydrogen and one quarter helium.

Galaxies at the limits of the observable Universe as seen by the Hubble Space Telescope. Their light has taken billions of years to arrive so we see these galaxies as they were billions of years ago. They have had billions of years less nuclear processing than our own Galaxy and therefore contain fewer heavy elements. (NASA/STScI)

The tiny proportion of 'everything else' is not uniformly distributed in the Universe. Instead, it is concentrated in particular places, such as on planet Earth. The distribution of elements is also very uneven. Fortunately, Earth has a sufficient quantity of elements such as carbon, oxygen and iron, to make life possible.

It seems that some kinds of atoms have scored much higher in Nature's popularity stakes than others. What is it about carbon or iron that makes them favoured by Nature in this way, and gold so rare? These questions can now be answered by bringing together knowledge from nuclear physics, astrophysics and cosmology.

Nuclei for life

The hydrogen nucleus usually comprises just a single proton, but it occasionally has one or even two additional neutrons, but never more. This limiting factor on the number of neutrons any nucleus can contain has a profound influence on the world of matter. If a nucleus of hydrogen had one proton and 99 neutrons, it would have very strange physical properties – for instance, water made with this isotope would be solid at room temperatures.

The carbon compounds upon which life depends would be quite different if ^{40}C rather than ^{12}C were the most common isotope of carbon; the fats in our bodies would weigh about three times as much and, even more dramatically, we would have great difficulty breathing out solid carbon dioxide! But ^{40}C does not, and cannot, exist. The key to understanding why lies in the energy stored within nuclei. Not only does this energy provide clues as to why certain nuclei are found in abundance and others are rare, but it also explains how energy is extracted from nuclei, both in the stars and on Earth.

What nuclei are possible?

Nuclei can be classified into two types: stable and radioactive. A stable nucleus lasts forever, but a radioactive one is unstable and decays into a different species. If it decays to another unstable nucleus, then that will also decay. Radioactive decay will continue until a stable nucleus is formed.

The presently known nuclei displayed according to proton number Z on the vertical scale, and neutron number N on the horizontal scale. The black squares denote the stable nuclei, and also the extremely long lived nuclei like ^{238}U that exist on Earth. The blue squares indicate nuclei with an excess of neutrons and so beta decay by emitting electrons. The red squares indicate nuclei which undergo positron decay (or electron capture). The yellow nuclei are those which decay by emitting alpha particles and the green nuclei undergo spontaneous fission. There are a few orange nuclei along the upper edge of the coloured area; these decay by emitting protons. The squares at the top right are recently produced super-heavy nuclei. This sort of diagram is called a Segrè chart.

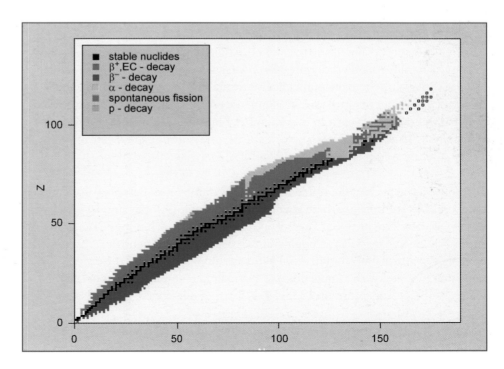

stable nuclides
β^+,EC - decay
β^- - decay
α - decay
spontaneous fission
p - decay

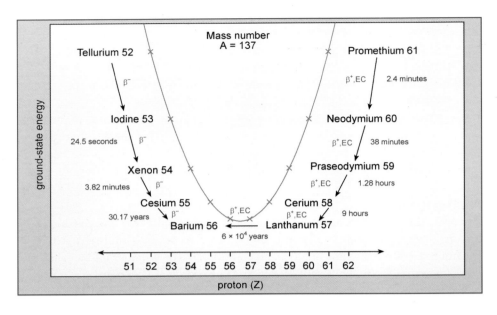

Mass number A = 137

Tellurium 52 ..... β^- ..... Promethium 61 ..... β^+,EC ..... 2.4 minutes

Iodine 53 ..... 24.5 seconds ..... β^- ..... Neodymium 60 ..... β^+,EC ..... 38 minutes

Xenon 54 ..... 3.82 minutes ..... β^- ..... Praseodymium 59 ..... β^+,EC ..... 1.28 hours

Cesium 55 ..... 30.17 years ..... β^- ..... Cerium 58 ..... β^+,EC ..... 9 hours

Barium 56 ..... β^+,EC ..... Lanthanum 57 ..... β^+,EC

6 × 10⁴ years

51 52 53 54 55 56 57 58 59 60 61 62

ground-state energy

proton (Z)

The unstable nuclei, those indicated with red or blue squares in the previous diagram, have the wrong proportion of protons and neutrons. They correct this by beta decaying to another nucleus with the same atomic mass number. The wrong balance of protons and neutrons gives the nucleus too much energy. This energy is carried off by the electron or positron and the neutrinos. This process of losing excess energy is rather like falling down a hill and this diagram shows how the energy increases rapidly away from the stable nucleus, in this case the barium isotope ^{137}Ba (the 56 indicates that barium is the element with 56 protons). This picture shows how the energies of series of nuclei, all with 137 nucleons, form a shape known as a parabola. The nuclei up the sides of the parabola eventually lose energy, falling down to the bottom in a series of beta decays. The half-lives for the decays are shown.

There are many more radioactive nuclei than stable ones. Some nuclei that have too many neutrons to be stable experience beta decay in which a neutron turns into a proton, emitting an electron and a ghostly neutrino. Nuclei with too few neutrons for stability undergo the other kind of beta decay in which a proton turns into a neutron.

All the possible stable nuclei have been known for many years, but the tally of known radioactive nuclei is continuously expanding. New and interesting radioactive nuclei are regularly produced in laboratories, enriching both our knowledge of nuclei and also helping us understand the workings of stars. Many of these new nuclei are very unstable, having half-lives much less than a second. Some survive for such a brief period, they can hardly be said to exist at all.

It is not yet known exactly how many or how few neutrons can exist in a nucleus with a given number of protons before it becomes so unstable that it effectively does not exist. So it is not yet feasible to predict the total possible number of nuclei.

The nuclei of most elements up to calcium, which has an atomic number, Z, of 20, have a handful of different isotopes, and in each case there is always a stable isotope with either the same number of neutrons as protons, or one additional neutron. The elements with an odd number of protons generally have fewer isotopes than elements with an even number. Fluorine, sodium and aluminium (elements 9, 11 and 13), for example, have just one isotope each, while oxygen, neon and magnesium (elements 8, 10 and 12), each have three isotopes.

For elements up to calcium, stable isotopes have similar numbers of protons and neutrons. Beyond calcium there is a change, and stable isotopes start to exist with more neutrons than protons. The isotope of calcium, ^{40}Ca, having 20 neutrons, is the last stable nucleus with an equal number of protons and neutrons. For heavier elements, stable nuclei contain more neutrons than protons and the excess of neutrons increases with atomic number. For example, stable isotopes of lead, which has 82 protons, contain 122, 124, 125 or 126 neutrons.

Some nuclei are not completely stable, but have long enough half-lives to be found naturally on Earth. Examples are the uranium isotope ^{238}U and the potassium isotope ^{40}K. The small amounts of ^{40}K that we all have in our bodies makes us slightly radioactive. This radioactivity contributes to the background radiation that we all experience. Half of all the ^{238}U that was present when the Earth was formed, 4.5 billion years ago, still exists in the soil and rocks. In fact, ^{238}U is much more common on Earth than gold, which is a stable element.

The driplines

There are some nuclei that take even more drastic steps to get their proton–neutron balance right – they emit protons directly. These nuclei, on the very edge of stability, lie along the 'proton dripline', marking the limit of the number of protons a nucleus can have, even for a very short time. Similarly, the neutron dripline marks the limiting number of neutrons a nucleus can hold without them being emitted immediately. At the time of writing the neutron dripline has only been reached up to about $Z = 12$, since the maximum possible number of neutrons is not yet known for larger nuclei.

For nuclei with more than 60 protons, there are other forms of radioactivity. Some decay by emitting alpha particles, others undergo spontaneous fission: they split, without any prompting, into two, approximately equal, lighter nuclei. This process has had momentous implications for humans and also, together with alpha decay, puts a limit on the total number of protons and neutrons that a nucleus can have.

The energy valley

Generally, the greater the imbalance of protons and neutrons in a nucleus, the shorter its half-life. When a nucleus undergoes beta decay, the emitted electron and neutrino carry off energy. The energy lost in this way is nearly always larger for decays between pairs of nuclei that are highly unstable. The more energy that is lost by the nucleus, the more rapidly the emission takes place and thus the shorter the half-life.

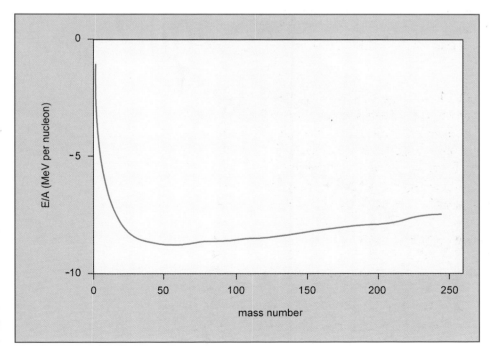

The total energy of a nucleus is not exactly proportional to the mass number, A, the number of nucleons. This graph shows how the energy per nucleon, written E/A, depends on A. The most stable nuclei are those near the lowest point on the curve, around mass number 56 to 60. Both the lightest nuclei and the heaviest nuclei have more energy for each nucleon than the nuclei with about 60 nucleons. That is why energy is released when the lightest nuclei undergo fusion to make heavier nuclei, or the heaviest nuclei undergo fission and split into two lighter nuclei. On this graph, the energy is that of the most stable nuclei with that particular mass number A, in other words the nucleus near the bottom of the parabola in the previous figure.

Among nuclei with a certain number of nucleons, the stable nuclei have the least energy while unstable nuclei have excess energy which they lose by transforming themselves through beta decay. The more unstable the nucleus, the more excess energy it possesses. It is as if the stable nuclei are at the bottom of a valley with the height above the valley floor being a measure of the energy stored in the nucleus. Nuclear physicists refer to this as the valley of stability. Continuing the valley analogy, nuclei with more energy are like boulders perched up the side of the valley; they are less stable than ones at the bottom – a nudge could send them rolling down. Not only does the valley slope upwards, but it also gets steeper further from the bottom. This is why beta decay, in which nuclei lose energy by jumping down the sides of the valley, takes place more rapidly for the distant nuclei – the downward jumps are greater. Boulders lying at the bottom of the valley need to have energy supplied to them to lift them up the side of the valley.

Along the valley floor

The amount of energy stored in a nucleus depends on the total number of protons and neutrons (the atomic weight A) as well as on the proportion of protons to neutrons. The more protons and neutrons in a nucleus, the more energy it has. The important quantity is really the energy per nucleon. This is just the total energy divided by A.

How the energy of a nucleus depends on Z, the number of protons, and N, the number of neutrons. The grey surface is in the form of a valley and the height represents the energy per nucleon, E/A. The valley floor follows the shape of the the curve on the previous figure, and the sides of the valley curve upwards as shown for mass number 137 in the figure before that. This figure also shows grooves or 'gullies' crossing the valley — these show the way the energy dips down for the magic numbers. The black pillars are there to guide the eye to the red lines which indicate the magic numbers. One gully crosses the valley for proton number 50, and another for neutron number 82.

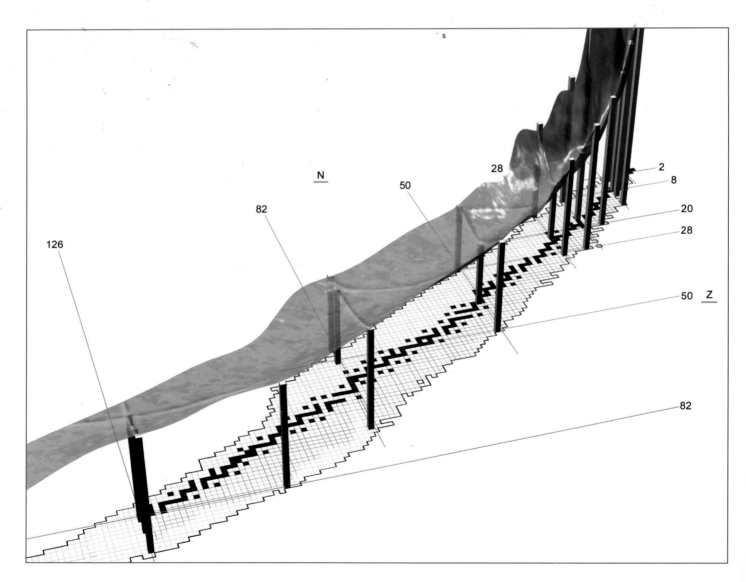

When all known nuclei are placed in a figure in which the height represents the energy per nucleon, the stable nuclei all lie along the floor of a valley. This floor is not level, but slopes toward a lowest point that is near particular nuclei with 26 or 28 protons. These nuclei have less energy per nucleon than any other nuclei. The bottom of the valley itself slopes upward slowly in the direction of the heaviest nuclei such as uranium and more steeply toward the lightest nuclei.

Element 26 is iron, and the position of the nuclei of certain isotopes of iron near the bottom of the valley is one reason why iron is very abundant both on Earth (the central core of the Earth is largely iron) and in the rest of the Solar System. Once the nuclei of iron are formed within stars, they are reluctant to transform into anything else without further energy being supplied. Elements with more protons than iron or nickel (element 28) tend to be less abundant on Earth and in the stars than most of the lighter elements. This is due largely to the way elements are made in stars, but the energy valley is a key factor.

In a similar manner to a river flowing along a valley floor towards the lowest point, the lightest atomic nuclei undergoing reactions tend to move towards the region near iron where the energy is lowest, by acquiring more protons and neutrons. Heavy nuclei can do this, somehow transforming themselves into nuclei with fewer protons and neutrons and thereby moving in the other direction. Thus two very different processes are involved, depending on the direction of the flow: nuclear fusion and nuclear fission.

Nuclear fusion and fission

If two very light nuclei can be made to coalesce or 'fuse' into one heavier nucleus, energy will be released because the heavier nucleus has less energy per nucleon than the lighter ones. Energy will therefore be 'left over' after the nuclei have coalesced. This is nuclear fusion, and is the source of the energy from the Sun. Most of this comes from hydrogen fusing to form helium. Fusion reactions that produce energy cannot produce nuclei heavier than iron or nickel, the nuclei at the lowest point of the valley.

In stars hotter and more massive than the Sun, other fusion processes occur. When a magnesium nucleus with 12 protons fuses with a silicon nucleus with 14 protons to make an iron nucleus with 26 protons, a great deal of energy is released. Much of the iron in our Solar System was probably made in processes like this when massive stars exploded more than four billion years ago.

The valley floor sloping upwards from iron in the direction of the heaviest nuclei means the amount of energy per nucleon is getting progressively greater. If a heavy nucleus splits into two, the resultant, lighter nuclei would exist lower down the energy valley and each would have less energy per nucleon. The total amount of energy contained by the resultant nuclei is less than the energy of the original nucleus. Therefore, after the split there would be some surplus energy. Such nuclear splitting is known as nuclear fission. When a heavy nucleus undergoes fission, the surplus energy is released mostly in the energy of motion of the two lighter fragments as they fly apart.

Fission is the source of much of the electrical energy used by many nations and it is the source of the man-made element, technetium, used to treat cancer and other diseases. In addition it forms the basis of weapons of dreadful destruction.

One of Nature's greatest surprises

The discovery of nuclear fission in the late 1930s came as a huge surprise. There was nothing in the behaviour of atoms to suggest that their nuclei should behave in such a way. Ernest Rutherford, and all the other nuclear pioneers, had given up hope that the energy locked within nuclei could be extracted and harnessed. Then, Otto Hahn and Fritz Strassmann were astonished to discover that when uranium was bombarded with slow neutrons, traces appeared of the much lighter element, barium. They worked hard to convince themselves that it really was barium produced, and not radium, an element with similar chemical properties but much closer to uranium in mass.

With hindsight, it might seem obvious that a uranium nucleus had split into two fragments, one of these being a barium nucleus, but this was initially far from obvious. It was only some months later, while on a skiing holiday in Kungälv in Sweden, that Lise Meitner and her nephew Otto Frisch came up with the now accepted picture of nuclei undergoing what they called fission – the word used by biologists for the process by which living cells divide. Frisch soon observed that the barium nuclei flew off at high speed, carrying away a lot of energy. The energy of nuclei could, after all, be extracted and maybe put to use.

When a nucleus undergoes fission, the two lighter nuclei that fly apart are soon slowed down by nearby atoms, to which some of the energy is transferred. In this way, a great deal of heat is generated which can be made to turn water into steam, powering turbines that generate electricity – but there is an unfortunate consequence.

Consider, once more, the bottom of the energy valley. For heavier nuclei, the stable isotopes contain more neutrons than protons, thus the floor of the valley curves away from the diagonal line that represents nuclei with equal numbers of protons and neutrons. If, for example, the uranium isotope ^{235}U absorbs a neutron, the resulting ^{236}U will undergo fission. It can split in a variety of ways, but will often split into nuclei of barium and krypton. These might be ^{146}Ba and ^{90}Kr; since barium has $Z=56$ and krypton has $Z=36$, the number of protons adds up to 92 – the correct Z for uranium. In addition, the total number of nucleons adds up to 236.

The problem is that the heaviest stable barium nucleus is ^{138}Ba and the heaviest stable krypton nucleus is ^{86}Kr, so both of the nuclei that are produced in fission have a large excess of neutrons. This means they are positioned somewhere up the side of the energy valley, an unstable state from which they will inevitably tumble down in a chain of beta decay steps.

Due to the curve in the energy valley, nuclei produced in fission will always be somewhere high up the neutron-rich side of the valley. The nuclei produced in nuclear fission are therefore highly radioactive, casting a shadow over applications of nuclear fission.

In fact, the fission products of a splitting uranium nucleus will not be quite so far up the valley walls as the example above suggests. When a nucleus divides in two, not all of its neutrons go into the two lighter nuclei. A few, generally two or three, are released. These can then be absorbed by other ^{235}U nuclei, turning them into fissionable ^{236}U. This is of great importance, making nuclear fission a practical source of energy, because these neutrons will then induce more uranium nuclei to

Lise Meitner, 1878–1968. Born in Vienna, she worked for many years in Berlin doing research in radioactivity. Element 109, meitnerium, is named after her. (AIP Emilio Segrè Visual Archives.)

Meitner and her nephew Otto Robert Frisch, 1904–1979, were the first to recognize a certain phenomenon discovered by Hahn and Strassmann as a signature of nuclear fission, the process shown above. Their fission model was based on Niels Bohr's 'liquid drop' model of nuclei, and predicted that fission releases much energy. At that time, in 1939, Meitner was taking refuge in Sweden. She spent her winter holidays with Frisch in the house on the right.

split. These in turn release more neutrons, each of which would cause more uranium nuclei to fission, and so on. This is the famous fission chain reaction.

When the chain reaction takes place in a slow and controlled way, we have nuclear reactors that can produce electricity and isotopes for medical purposes. When it happens in an uncontrolled way, we have a nuclear bomb.

Limitations of mass

In about 1942 it was found that uranium nuclei can, very occasionally, undergo fission without any stimulation by neutrons. This process is called spontaneous fission and helps answer the question of which nuclei can exist. Spontaneous fission becomes more and more important for the very heaviest nuclei, and is one of the factors that determines just how heavy a nucleus can be.

At the top end of the valley of stability, most nuclei decay by emitting alpha particles. Alpha particle decay can also be thought of as a kind of fission in which one of the nuclei produced is very much smaller than the other. Energy is released, and carried off by the alpha particle, in a similar manner to when a nucleus undergoes fission. The resulting nucleus lies further down the energy valley than the original one and, although the alpha particle itself is far down at the other end of the valley floor, it is also a very stable nucleus.

The tendency for nuclei to alpha decay is also a major limiting factor to how heavy nuclei can be. The half-life of a nucleus undergoing alpha decay is exceptionally sensitive to the energy of the alpha particle: the more energy carried off by the alpha particle, the shorter the half-life. Thus the excess energies of the alphas at the extreme top end of the valley indicate that these nuclei are very short-lived.

Magic numbers

So far, the energy valley has been presented as being smooth, but studied more closely, it is seen to be rather bumpy. This is because the energy per nucleon of stable nuclei does not follow a smooth curve. A select few are slightly lower than their neighbours since they have even less energy per nucleon (more stable). Such nuclei tend to line up together in grooves that criss-cross the valley floor. These grooves occur for particular numbers of protons and neutrons known as magic numbers. A good example occurs with the series of nuclei lying on the line for 50 protons. The isotopes of the element with $Z = 50$, tin, have somewhat lower energy per nucleon than their neighbours, and have various other properties that are associated with greater stability. Tin also has more stable isotopes (10 in all) than any other element.

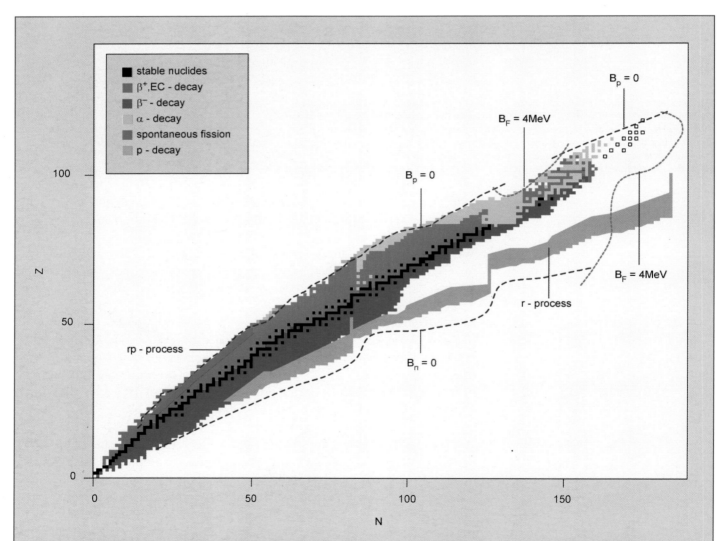

When viewed on a Segrè chart like this, the possible nuclei do not extend indefinitely in all directions. The limits are called the 'driplines' and are shown in this figure. For example, add a neutron to a nucleus on the neutron dripline (B_n), and it would 'drip' right out again. Apart from the lightest nuclei, we do not get near the neutron dripline. We do reach the proton dripline (B_p) in some places. The possibility of spontaneous fission also imposes a limit on which nuclei can exist; the green line marks this limit. Beyond it, nuclei are expected to live so fleetingly that they can hardly be said to exist. Later on we explain the r-process (pink areas) and the rp-process (green).

The number 50 is also significant for neutrons. All the nuclei on the line signifying 50 neutrons have a little less energy per nucleon than their neighbours. Thus a nucleus with either 50 protons or 50 neutrons will be significantly more stable than nuclei with slightly fewer, or more, protons or neutrons. Although they are not deep compared to the main energy valley itself, these grooves have a profound effect on what nuclei exist, and also contain vital clues about the structure of nuclei.

There are other magic numbers. The complete set is: 2, 8, 20, 28, 50, 82 and 126.

Nuclei that have magic numbers of both protons and neutrons are especially stable. The simplest example is ^{4}He, the alpha particle, with two protons and two neutrons. This is the reason for its exceptional stability. Another example is the isotope of lead, ^{208}Pb, with 82 protons and 126 neutrons. Lead is common on Earth, and ^{208}Pb is its most abundant isotope. However, lead is surrounded by quite rare elements that have relatively unstable nuclei. The next element after lead is bismuth, Z=83, which is hundreds of times rarer than lead in the Earth's crust. All elements with more than 83 protons only have radioactive isotopes.

The possible nuclei

The energy valley answers some questions about what nuclei exist; for example, it explains why there are no ^{40}C or ^{100}H nuclei – they would lie too far from the bottom of the energy valley. The existence of a nucleus containing a particular combination of protons and neutrons depends on the energy of the combination. Some combinations yield nuclei that last forever, others are subject to radioactive decay. In general, the further a nucleus is from the valley floor, the shorter its half-life.

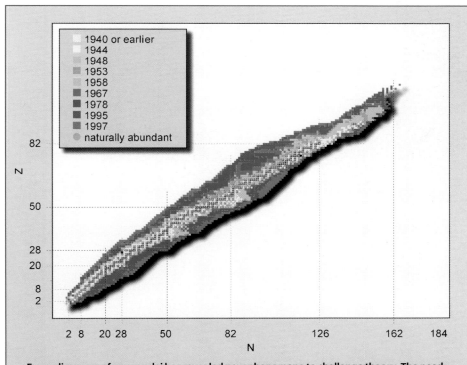

Every discovery of new nuclei has revealed new phenomena to challenge theory. The need to understand nuclear processes in stars makes it even more urgent to push further into the unknown, since many nuclei near the neutron dripline play key roles in the processes that form the heavy elements. Here we see how active this adventure into dripline territory has been in recent years.

Too far from the valley floor and any protons or neutrons that we might try to add to a nucleus leak right out again. These points mark the boundaries of the region where nuclei either exist or would exist if we could make them. These boundaries are the driplines mentioned earlier. In some places we do know where they are, and in other places we can make informed guesses.

There are many nuclei that must exist but which have not yet been identified in the laboratory. The number of nuclei that have been created and studied increases year by year.

The super-heavies

Although all the isotopes of thorium, Z=90, and uranium, Z=92, are strictly unstable, both elements have one or two isotopes with extremely long half-lives – as long as the age of the Earth in the case of uranium, and three times longer than that for thorium. All isotopes of elements with more than 92 protons are much too short-lived to have survived since the Solar System was formed. It naturally became an irresistible challenge to make nuclei with more than 92 protons using nuclear reactions! These are known as the transuranic nuclei since they have more protons than uranium.

Element 93, neptunium, was identified by Edwin McMillan and Philip Abelson in Berkeley, California, in 1940. Over the years, elements 94 (plutonium) to 103 (lawrencium) were also produced at Berkeley as a result of heroic efforts by Glenn Seaborg, Albert Ghiorso and their large team using cyclotrons and other accelerators. Elements 104 (rutherfordium) to 107 were for many years the subject of claims by workers at Berkeley and at Dubna in Russia.

One of the first large cyclotrons at the Lawrence Radiation Laboratory, Berkeley, 1944. From left, Luis Alvarez, William Coolidge, William Brobek, Donald Cooksey, Edwin McMillan and Ernest Lawrence . Alvarez and Lawrence were later awarded Nobel prizes in Physics, Lawrence for inventing the cyclotron and Alvarez for discovering many new particles using bubble chambers. McMillan, discoverer of element 93, neptunium, shared a Nobel prize in chemistry with Glenn Seaborg for their work on transuranic elements. McMillan later helped develop the synchrotron, the high energy successor to the cyclotron. (Berkeley National Laboratory, AIP Emilio Segrè Visual Archives.)

Advances in making new elements are currently being made at GSI near Darmstadt in Germany. The challenge the physicists face is formidable. As the number of protons increases, the half-lives of these elements gets shorter. For example, the longest lived isotope of rutherfordium, ^{261}Rf, has a half-life of about one minute. Alpha decay and spontaneous fission compete to make nuclei of the other isotopes vanish even more quickly. Moving up to element 107, half-lives are typically measured in milliseconds, but the ambition to reach higher atomic numbers remains.

For many years, nuclear theory has suggested that 114 is a new magic number for protons. This might lead to an 'island of stability' around elements 112 or 114, where lifetimes are longer than expected and perhaps become long enough for the chemical properties of the elements to be studied. The term 'island' is a little misleading since it implies a higher level than its surroundings – in the picture of the energy valley, these more stable heavy nuclei would actually reside in a lower energy region than their surroundings. It is not yet clear whether this island of super-heavy nuclei exists.

The curvature of the energy valley adds to the difficulty of making super-heavy nuclei. It dictates that the heavier the nucleus, the greater the excess of neutrons over protons, but super-heavy nuclei must necessarily be made using reactions with lighter nuclei, having a smaller ratio of neutrons to protons. It follows that the nuclei made in these reactions will have too few neutrons to be near the bottom of the energy valley. To make the job even harder, these heavy nuclei undergo fission with little prompting, and the very reactions in which they are made are likely to put a lot of energy into the system. Too much energy would nudge any nucleus with a tendency to fission to take the plunge and break up again.

Even if a few super-heavy nuclei are made, identifying them is extremely hard. Finding a needle in a haystack is nothing compared with having to identify the single super-heavy nucleus produced per day when trillions of particles are hitting the detectors every second. The efforts to study these exotic nuclei are very rewarding to both physicists and chemists; chemists can check whether they can predict the chemical properties of hitherto unknown elements, while physicists have an invaluable test of their theories, which predict the probability of alpha decay and spontaneous fission.

Step by step the GSI physicists have worked their way up through new elements with shorter and shorter half-lives. These include element 109, meitnerium (named after Lise Meitner), identified in 1982, and the yet-unnamed elements 110 and 111 discovered in 1994, and 112 discovered in 1996. There have been reports that element 114 might have been discovered in Dubna in 1999, and most recently there is evidence from Dubna for element 116.

Exotic nuclei at the driplines

Great efforts are being made to extend our knowledge of nuclei in the directions of the lowest and highest possible neutron numbers for a given Z – the most neutron-rich and neutron-deficient isotopes of an element. Such unstable nuclei are hard to make on Earth, but they do exist briefly when stars explode. They are created en route to producing many of the elements on Earth, including some of those inside our bodies. We can never fully understand where the elements on Earth come from, or the stars they were made in, without studying the exotic, highly unstable nuclei closely.

Glenn Seaborg, 1912–1999, discoverer of plutonium and other elements. Element 106, seaborgium, is named after him.

7. Applications of nuclear physics

Interactions with everyday life

By the 1930s, nuclear physics was entering a golden age. Almost all the elements were known and their chemical properties established: from hydrogen with its nucleus consisting of just one proton, to uranium with 92 protons and well over a hundred neutrons. There remained just a few conspicuous gaps in the periodic table – apparently, elements with certain atomic numbers simply did not exist on Earth. Nuclear physicists have since artificially created and identified the nuclei of these elements; one, in particular, has turned out to play a vital role in modern medicine and has also become of great significance to astronomers.

The element in question has atomic number 43 and was discovered in 1937. In that year Ernest Lawrence used his new particle accelerator, the cyclotron, to send a beam of deuterons onto a target containing element 42, molybdenum. When the deuterons, the nuclei of an isotope of hydrogen containing just a proton and a neutron bound together, interacted with the molybdenum nuclei, reactions took place making the target radioactive. Some deuterons deposited their neutron in the molybdenum target nuclei, while their proton continued on its way; the extra neutron in these molybdenum nuclei caused them to become unstable and suffer beta decay, with a neutron becoming a proton accompanied by the emission of an electron and an anti-neutrino. In this manner, element 42 becomes element 43.

Lawrence did not study the molybdenum targets himself – he sent them off to a promising young student of Fermi, Emilio Segrè, who had just taken up a post in Palermo, Sicily. Segrè and his colleague, Carlo Perrier, analysed the targets and found traces of a hitherto unknown element which they succeeded in identifying as the missing element 43. Later, they called this element technetium, since it was the first element produced by artificial, or 'technological' means. It was given the symbol Tc and a gap in the table of the elements was filled.

No one had found technetium on Earth because it has no stable isotopes. Even the longest-lived isotope has a half-life of just over four million years – short enough for any amount of technetium originally present on Earth when the Solar System was formed to have decayed away.

Subsequently a number of other missing elements – 61 (promethium), 85 (astatine) and 87 (francium) – were also produced using nuclear reactions, but it is technetium which has become deeply significant to the lives of many thousands of people. As an example of its importance, when the world supply of technetium was threatened by a strike at a nuclear laboratory in Canada in 1998, the President of the American College of Nuclear Physicians wrote a strongly worded letter to the Canadian Prime Minister pointing out that 47,000 medical procedures per day, in the USA alone, were under threat. In addition, medical procedures throughout the world were being cancelled.

Emilio Segrè,1905–1989, shared the 1959 Nobel prize in physics with Owen Chamberlain for the discovery of the anti-proton. He had many other discoveries to his credit, including new elements. (AIP Emilio Segrè Visual Archives)

The technetium story is one example of how human exploration of the heart of matter has led to discoveries which have become part of everyday life. Modern medicine and industry would be untenable without them.

Detecting and measuring radiation

Before exploring how radioactive elements are used in medicine and industry, we must examine what happens when radiation passes through matter and how it is detected and measured.

The key to all the effects of radiation – useful and life saving, as well as damaging and life threatening – is its power to ionize the material through which it passes. Alpha, beta and gamma rays are all examples of ionizing radiation. As an alpha particle, for instance, passes through matter, it knocks electrons out of atoms and molecules, ionizing them. This can destroy the molecules and, if they are in living tissue, temporarily or permanently damage the cells containing them. As it ionizes matter, the alpha particle will lose energy and slow down until it eventually comes to rest.

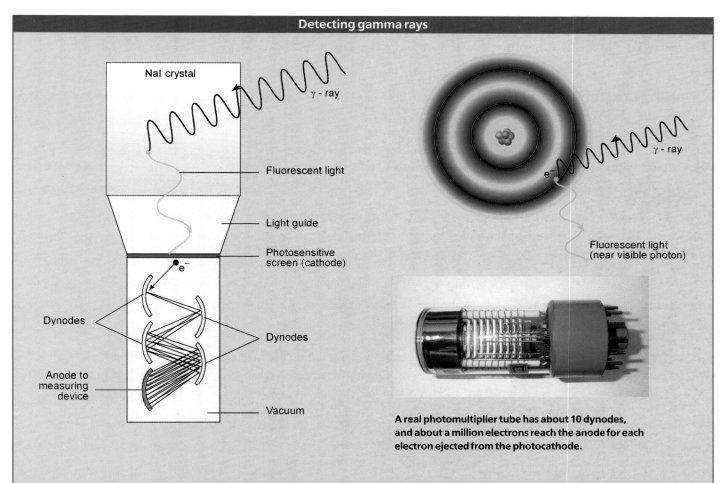

Detecting gamma rays

NaI crystal

γ - ray

Fluorescent light

Light guide

Photosensitive screen (cathode)

e⁻

Dynodes

Dynodes

Anode to measuring device

Vacuum

γ - ray

e⁻

Fluorescent light (near visible photon)

A real photomultiplier tube has about 10 dynodes, and about a million electrons reach the anode for each electron ejected from the photocathode.

A modern means of detecting gamma rays and measuring their energy relies on scintillations, the flashes of light given off as gamma ray photons interact with suitable substances. A common choice is a crystal of sodium iodide, NaI, doped with about 1% of thallium. A gamma ray disturbs the electrons of many atoms (one is shown here) and each atom emits photons of visible light as it settles down. The higher the energy of the gamma ray, the more photons are emitted. By measuring the intensity of the flash, the energy of the gamma ray can be measured. Photons comprising the flash are guided to a photomultiplier where they release electrons from a photocathode into a vacuum tube containing a series of dynodes. The electrons are accelerated from dynode to dynode by an electric field, and each time they strike a dynode more electrons join the stream, ending up at the anode where they enter an electronic measuring circuit. The whole system is called a gamma ray spectrometer.

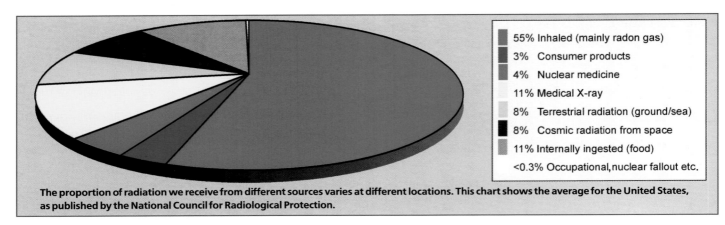

The proportion of radiation we receive from different sources varies at different locations. This chart shows the average for the United States, as published by the National Council for Radiological Protection.

Gamma rays and X-rays also ionize the material they pass through. They eventually disappear by being absorbed by the atoms, rather than losing energy by slowing down. As they are forms of electromagnetic radiation, there is only one possible speed for X-ray and gamma ray photons: the speed of light.

All kinds of ionizing radiation have a range beyond which they are effectively fully absorbed. This range depends on the energy and the nature of the material; lead, for example, is more effective than most materials at stopping radiation and is therefore used for shielding.

Natural and artificial radiation

Natural radiation is all around us. Over the history of life on Earth, living creatures have evolved senses which warn of dangers such as excessive temperature and approaching predators, but humans, at least, have no sense that responds rapidly to ionizing radiation. We are quite unaware of any variation in the radiation to which we are subjected. Natural sources of radiation include the ground we walk on, the air we breathe and the food we eat. Some naturally occurring radioactive isotopes, such as uranium and thorium, are left over from the time when the Earth was formed – others, such as the carbon isotope ^{14}C, and tritium, ^{3}H, are made by interactions within the atmosphere when high energy particles (cosmic rays) bombard the Earth from space.

One of the most abundant radioactive isotopes found on Earth is that of potassium, ^{40}K, which makes its way into the food chain. It exists as small traces alongside the stable isotope, ^{39}K, which is vital for life. Once ingested, the traces of ^{40}K decay inside our bodies. Another radioactive element absorbed by humans is ^{14}C. All living organisms take in carbon, which contains a constant fraction of radioactive ^{14}C. Once an organism dies and stops taking in carbon, the amount of ^{14}C can be used to determine the time elapsed since death. This is the basis of carbon dating.

The largest amount of natural radiation to which humans are exposed is by breathing in the noble gas, radon. Radon is a decay product of uranium found in rocks. It is chemically inactive allowing it to migrate through porous materials such as house foundations. Once in the air inside the house it can be breathed in, where it has a chance of decaying inside the lungs.

The amount of radiation we receive depends on the kind of rocks in the ground beneath us, how high we live above sea level and, in recent times, what sort of job we do. Dentists giving X-rays to their patients, or miners working underground in certain kinds of rock, will receive extra radiation. Airline crews who spend more time than most at high altitudes also receive extra radiation from the effects of cosmic rays. The natural background radiation varies a great deal, and is usually much greater than man-made radiation.

We all experience some man-made ionizing radiation – from minute amounts emitted by smoke detectors to sometimes quite substantial doses encountered in medical treatments. Many people have jobs involving low levels of radiation and we also receive radiation from the nuclear power industry. While these sources are usually very much smaller than the background, the Chernobyl incident serves as a warning that this powerful force must be treated with respect.

Nuclear physics in medicine

Everybody will know of someone who has benefited directly from applications of nuclear physics in medicine. Some measure of its importance can be judged from the cold monetary figures: worldwide, nuclear medicine is a $10 billion per year business. In Europe and the US, radioisotopes are used in some way in the treatment of almost half of all patients admitted to hospital.

Nuclear medicine is important for both diagnosis and therapy. Diagnosis ranges from routine use of X-rays to injection of radioactive material for gamma imaging, while the best-known use of radiation for treating disease is radiotherapy used against cancer. The power of radiation to cure is often less publicized than its power to do harm – radiation as invisible scalpel rather than invisible dagger. Just as a scalpel must be wielded with extreme care, so radiation must be carefully controlled. Doctors who prescribe such radiotherapy must weigh the risks of radiation against the benefit of the treatment.

The effect of radiation on living tissue does not just depend on how much energy is deposited, but also on the type of particles involved. When a gamma ray strikes the nucleus of a cell, one half of the double strand of the genetic material DNA might be broken, but there is a good chance that the cell will repair itself. When an alpha particle passes through a cell, the much more intense trail of ionization is more likely to destroy both strands, something the cell cannot repair. Fortunately, alpha particles have a short range, and will not pass through human skin; unfortunately, there are some sources of alpha particles, such as radon gas, which can enter the body just from the air we breathe. However, extensive surveys in the US and UK have found virtually no correlation between environmental radon levels and disease, which illustrates the adaptive power of living organisms.

There are two reasons why radiation from radioactive nuclei and high energy particles from accelerators are powerful weapons against disease. Firstly, radiation preferentially destroys fast growing tissues such as cancers. Secondly, radiation can be concentrated in the specific tissues which need to be destroyed. This is particularly true for beams of protons and heavier nuclei. Such particles give up most of their energy at the end of their path so that it is possible to arrange for the dose to be delivered to a well-defined volume within the patient's body.

Before a medical condition can be treated, it must be detected and a diagnosis made. Many gamma-emitting radioisotopes, such as technetium, can be used as 'tracers'. Wherever a tracer goes in the body it can be followed by suitable detectors which measure its characteristic radiation. It can be made into a chemical form that concentrates in particular organs, allowing the precise location, shape and biochemical function of those organs to be mapped out. This is something X-rays cannot do with soft tissue.

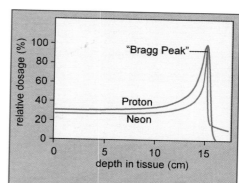

Charged particles travelling through matter give up most of their energy towards the end of their paths as they come to a stop. This is particularly true for heavier nuclei like those of neon, where the energy is sharply concentrated at a single point. For protons, the energy is deposited at a wider range of depths.

Using beams of carbon nuclei to destroy cancer; the first patient on the treatment couch at GSI, Darmstadt, Germany. (Gaby Otto, GSI.)

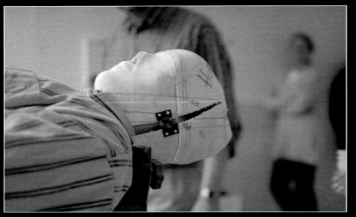

A patient prepared for therapy for a brain tumour using a beam of high energy carbon nuclei at GSI. (Achim Zschau, GSI.)

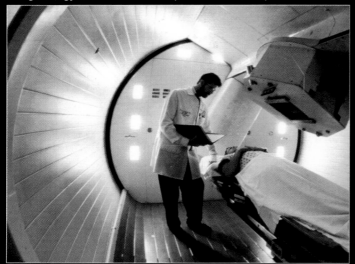

The neutron beam treatment room at Harper Hospital in Detroit. (Courtesy Harper Hospital, Detroit, Michigan, United States.)

Unlike X-rays, heavy ions deliver their energy towards the end of their path through tissue. At GSI in Darmstadt, Germany, accelerated beams of carbon nuclei concentrate their destructive power precisely throughout the volume of a cancer, with relatively little damage to the rest of the body. Neutron beams, used at Harper Hospital in Detroit are effective against particular cancers such as prostate cancer. Since they are uncharged, they cause no ionization along their path through the tissue, until they strike their target.

The magnetic pole-piece, an essential part of the neutron therapy cyclotron at Harper Hospital. (Courtesy Harper Hospital, Detroit, Michigan, United States.)

airline luggage for hidden explosives; and krypton, which is used in the indicator lights of many household appliances from washing machines to coffee makers. In addition, tritium, the isotope of hydrogen with two neutrons (as compared with deuterium which has one), is used to make self-luminous signs and paint, and is found in wrist watches for the same purpose.

Even uranium isotopes have many uses, such as in dental fixtures and wall tiles, while plutonium, the element most closely associated with nuclear weapons, has safely powered many spacecraft since 1972.

Nuclear physics in astronomy

Almost everything we see when we look up into the night sky is the result of nuclear reactions. Stars shine because nuclear reactions are taking place at their centres, but their nuclear fuel does not last forever. The detection of the short-lived element technetium in stars which are billions of years old proved such elements are created within them.

As nuclear fusion processes turn hydrogen into heavier elements, generating heat, stars undergo a series of changes at a rate which depends on their mass. The heaviest burn their nuclear fuel very quickly before exploding as supernovae. Lower mass stars like the Sun will, after shining for a few billion years, expand and cool. During this stage they are known as 'red giants', and when the Sun becomes one, its outer atmosphere will swallow the Earth.

Red giants also played a vital part in our history. The Solar System condensed out of gases and dust that was enriched by cinders from exploding stars and by material thrown off by an earlier generation of red giants. Many of the heavy elements existing on Earth, like the lead on our roofs, were produced by these red giants in the 's-process' which allows heavy elements to be formed by the slow addition of neutrons.

The constellation of Orion. The star at top left appears to have a distinctly reddish colour when seen with the naked eye. It is the red giant star Betelgeuse. Betelgeuse is large enough to contain the path of the Earth around the Sun. Our understanding of the processes taking place in such stars rests critically on what happens when nuclei collide.
(Courtesy Christopher Doherty.)

What elements are present on the surface of Mars? How much carbon, or any other element, is there in a biopsy sample and how is it distributed in the sample? Questions like these arise time and again in medicine, in industry and in all kinds of research including planetary exploration.

A very old idea, reborn with modern techniques, comes to the rescue. It harks back to the alpha particle scattering experiments which revealed to Rutherford that all atoms had central nuclei. In those experiments, alpha particles were fired at a sample and a few came straight back. Geiger and Marsden's targets were pure metals such as gold. But a sample might have many different elements in it. If so, the energy the alpha particles have when they fly back to be detected depends on the mass of the nucleus which they strike, just as a tennis ball bounces back with more energy if it hits a soccer ball rather than a ping-pong ball. So if we can measure the exact energy of each alpha particle which bounces straight back, and count how many have particular energies, we can say exactly how much of each element there is in the part of the sample we are studying. This way of studying samples is called Rutherford Backscattering Spectrometry (RBS). It was used to make the first measurements of what elements were on the Moon long before people went there. Alpha particles from an Americium, ^{241}Am, source, which was landed on the Moon, were bounced off the lunar surface. The alpha particles were then counted and sorted by energy and the numbers beamed back to Earth.

More recently, in 1997, RBS was employed by the Mars Pathfinder rover, named Sojourner, to send back to Earth information about the elements on the surface of Mars. In this case, it was part of a package of three methods which all made use of alpha particles emitted by the transuranic element Curium 244 (^{244}Cm). The lighter elements in the rocks, like carbon and oxygen, were measured by RBS. Somewhat heavier elements, like fluorine and sulphur, were analysed by nuclear reactions in which an alpha particle is absorbed by a nucleus and knocks out a proton. By measuring the energy of the proton, the target nucleus can be identified. The third technique is called PIXE (Particle Induced X-ray Emission). When the alpha particles strike somewhat heavier atoms, such as iron, they knock electrons out of their atomic orbits. As a result, the atom emits X-rays as they settle down after the disturbance. These X-rays allow the element to be identified just as Moseley discovered early in the 20th century. In this way, using the combined package of RBS, nuclear reactions and PIXE, Sojourner was able to send back a detailed account of what elements there were in the rocks it encountered on Mars. Earlier Viking missions to Mars had no means for measuring the lightest elements. Sojourner is shown on the next page.

We also find RBS and PIXE working together in medicine. For example, there are diseases caused by too much iron in cells. Can we see where iron might be concentrated within the cells? Ordinary chemical techniques could tell you how much iron was in a sample of tissue, but only by destroying the tissue — then you could never know how it was distributed within the cells. When working together, PIXE finds heavier elements, such as iron, in samples of tissues and RBS measures the lighter elements in the same samples, thus revealing the cell structure. By putting this information together, RBS and PIXE allow us to map out the excess iron within the cells. Very finely focussed beams of nuclear particles are needed to do this. In the case shown below, proton beams were used. In the left-hand picture we see the carbon intensity as revealed by RBS using a proton beam that can be narrowed down to a few millionths of a metre. This is a skin sample obtained by biopsy from a patient. The colour is coded with yellow indicating a greater concentration of carbon; the walls of a single cell can be seen. The picture on the right shows the location of iron as revealed by PIXE with the same proton beam. Too much iron in some cells has caused their membrane to burst, killing the cells. Liver cells are particularly susceptible. RBS and PIXE together allow doctors to study exactly what goes on when this fatal disease, called haemochromatosis, strikes, a necessary first step to finding a cure.

Today, RBS has a wide range of industrial applications. It is an ideal way of looking for defects in the very thin multi-layered structures found in modern micro-electronics. RBS can call upon the full range of modern nuclear physics techniques for detecting particles, measuring their energies and counting them. Particle accelerators can provide beams of whatever particle is most suitable, not just alpha particles, and direct them with great precision onto the part of the specimen to be studied.

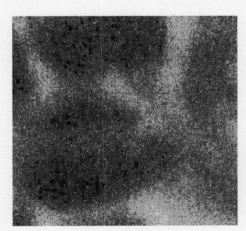

The yellow areas are the cell walls, where the carbon concentration is higher, as revealed by Rutherford backscattering using a fine beam of protons.

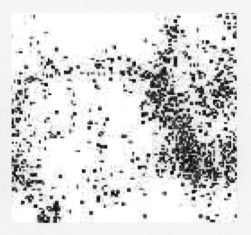

The red dots represent iron as revealed by PIXE using the same beam. Together, the figures show us where the carbon is within the cell. These two elemental maps were obtained at the Institute for Nuclear Technology, Lisbon, Portugal.

Inset into the wide field view of the Martian surface seen from the Mars Pathfinder lander is the rover, named Sojourner, which was equipped with an alpha–proton–X-ray spectrometer to analyse the elements in the Martian rocks. (Courtesy NASA/JPL/Peter Smith.)

8. Nuclear architecture
From hydrogen to neutron stars

A single remarkable pattern, the energy valley, sums up a great deal of information about nuclei, including which nuclei exist, which are stable, and how much energy can be extracted from them. To investigate the energy valley further, we must delve into the architecture of nuclei. The structure of nuclei depends on the quantum rules that control the properties of the protons and neutrons, and the nature of the forces between them. To follow the architectural analogy, the nucleons are the bricks, and the nuclear forces provide the mortar, while everything is under the control of rigorously enforced planning regulations provided by the quantum rules.

The force between protons and neutrons

The forces that hold nuclei together are not like any in the everyday world. Nuclear forces only act over extremely short distances – distances so short, that a nucleon on one side of a heavy nucleus, such as a lead nucleus with 208 nucleons, cannot feel the presence of a nucleon on the opposite side. The range of the force is about 2 femtometres, compared to 13 femtometres for the distance across a lead nucleus. The short range of the force between nucleons has a big influence on the shape of the energy valley.

The reason for the short range of the nuclear force can be understood in terms of the exchange of 'force carrying' particles between the two interacting nucleons. We have seen how a pion, created using borrowed energy, is exchanged between two nucleons to generate the force between them. The range has to be small since the pion can travel only a very short distance before Nature calls back the loan. Electric and magnetic forces have longer ranges since they result from the exchange of photons, which are massless. For photons, Nature allows an unlimited loan period.

The nuclear force is also very much stronger than the other forces in nature. This strength is partly what causes the energy bound up in nuclei to be about a million times greater than the energy holding atoms and molecules together. Consequently, about a million times as much nuclear energy can be extracted from a given mass of uranium than the chemical energy from the same mass of coal.

The force between a pair of nucleons is also highly complex, and this book will not examine it in depth. Its strength, for instance, depends on the way the two interacting nucleons are spinning relative to each other. It also depends on whether they are the same type of nucleon (two protons or two neutrons) or a proton and a neutron.

The force is not quite strong enough to bind two neutrons or two protons together, but is just strong enough to bind a proton and a neutron to make a deuteron. The force between two protons is the same as the force between two neutrons except that there is also the electric repulsion between them. This electric force is, how-

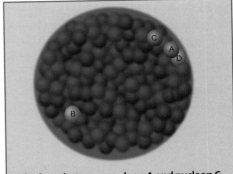

The force between nucleon A and nucleon C is strongly attractive. Nucleon D, however, is too close and is repelled by nucleon A, while nucleon B on the other side of the nucleus is outside the range of the force due to A.

ever, completely overwhelmed by the nuclear force when the protons are close together, but is significant for two protons on opposite sides of a large nucleus. Another feature of the force between two nucleons is that it changes from being attractive to repulsive at very short distances so that two nucleons cannot overlap (under normal conditions) in a nucleus.

The nuclear force is incredibly finely balanced. A few percent stronger and two protons (or two neutrons) would bind together. Nuclei consisting of just two protons do not exist, but if they did all the hydrogen would have been consumed in the Big Bang leaving none to power ordinary stars, including the Sun. It would then have been impossible for life to evolve on Earth. A few percent weaker and the deuteron would not be bound. Most of the energy from the Sun comes from a process which starts with the formation of deuterium, so if this occurred, again life would not have evolved on Earth. In either case, the elements produced in the later stages of the Big Bang would have been quite different.

Quantum rules rule

According to quantum theory, a proton or a neutron does not have a particular place in a nucleus in the same way that a brick has a fixed location in a house. Each proton and neutron has a probability density that is spread throughout the nucleus. There are more and less likely places for a nucleon to be found, as determined by its wave function. This is the mathematical rule specifying the probability for a nucleon to be found at any particular place. It describes an orbit, a little like the orbit of electrons in atoms and not at all like the orbits followed by planets. While it is difficult to visualize electrons in their smeared out orbits around nuclei, it is even harder to imagine how nucleons might move round in the crowded nucleus.

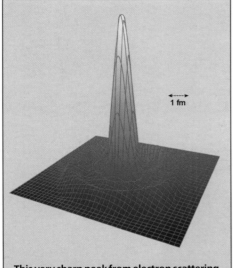

This very sharp peak from electron scattering experiments shows that the 82nd proton in lead is most likely to be found near the centre of the nucleus. The outer rings show that the neutron can also be found at about 4 and 7 femtometres from the centre of the nucleus. In most cases, the last proton in a nucleus does not have such a large probability of being at the centre of the nucleus.

1 fm

Nucleon orbits have one dramatic property: the less spread out in space an orbit is, the more energy of motion (technically, the kinetic energy) the nucleon has. This is one aspect of Heisenberg's Uncertainty Principle and is the main reason why nuclei do not collapse to a point even though the very strong nuclear force is pulling each nucleon towards its neighbours. The fact that the nuclear force between nucleons becomes repulsive at very short distances stops the nucleus being quite a point, but without the Uncertainty Principle nuclei would definitely be much smaller.

If a nucleus was smaller, each orbit would have to be more compact simply to describe the smaller volume in which a proton is found. The Uncertainty Principle imposes a cost for this to happen: in a smaller volume, the energy of motion becomes larger, which means energy needs to be added to a nucleus to make it smaller. As a result, nuclei are highly incompressible. The actual size of a nucleus with a particular number of nucleons is a balance between the nuclear forces pulling the nucleons together and the Uncertainty Principle keeping them apart.

Two other factors work together with the Uncertainty Principle to dictate nuclear sizes. One is the repulsive part of the force that takes over when nucleons get very close to each other. The other is the quantum rule known as the Pauli exclusion principle. This dictates that only two protons and two neutrons can occupy each possible quantum orbit.

Every orbit must obey the Uncertainty Principle, but some orbits are more compact than others. Nucleons fall preferentially into the innermost, most compact, unfilled orbits. Adding nucleons to a nucleus progressively fills up shells, making it bigger. The nucleus grows in such a way that its central density remains close to the characteristic density for the centre of all nuclei. In this respect, nuclei are very different from atoms, which are all roughly the same size. In atoms, all the electron orbits get smaller as electrons are added. This is because an atom will only be able to accommodate more electrons if its nucleus has more protons, and this extra positive charge draws all the electron orbits inwards. There is nothing equivalent at the centre of the nucleus to pull in the nucleon orbits.

Two contrasting pictures

Two quite different pictures of nuclei have been presented. On the one hand, protons and neutrons within nuclei lie in orbits much as electrons in atoms, but on the other hand nuclei behave in many ways as if they were made of incompressible fluid. All nuclei except the very lightest have the same density, much as water droplets all have the same density. If the nucleons are so closely packed together it is hard to see how they could possibly be orbiting around inside the nucleus at the same time. Water molecules certainly do not travel in simple orbits within a water droplet.

The energy valley reflects the two contrasting aspects. In the first place it has an overall smooth structure which can be summed up in just three curved lines: the height of the valley floor (below), the parabola on the following page, and the bend in the line of black squares in the Segrè chart (page 71) as N becomes greater than Z for the heavy stable nuclei. However, the basically smooth valley is crossed by a number of small ditches or gullies following the lines of the magic numbers. It turns out the magic numbers are a signature that nucleons lie in simple orbits.

Are nuclei, then, like water droplets or like miniature solar systems with nucleons in orbits? The answer is that they are not quite like either but are somewhat like both. This might seem paradoxical, but our expectations are formed by our experience with the world of visible things, and the quantum world is very different.

A droplet of nuclear material

Consider the way the floor of the energy valley first slopes down as we move from the lightest stable nuclei until it reaches its lowest point around iron and nickel, and then slowly rises again towards the heavy nuclei. This overall behaviour can be explained in terms of the average force felt by each nucleon.

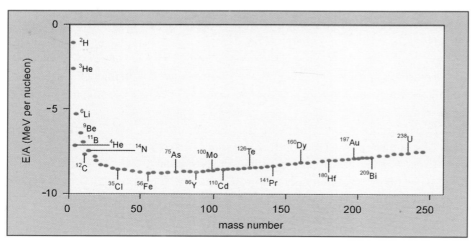

The floor of the energy valley showing where some particular nuclei lie. The height is the energy per nucleon, labelled E/A. It is lowest near iron (Fe), highest near the lightest nuclei such as hydrogen (H), and gradually becomes higher for elements heavier than iron.
The shape of the energy valley explains why energy is released by fusion of light nuclei and by fission of heavy nuclei.

Nucleons deep within a nucleus 'feel' the attractive force of nucleons surrounding them, as if they have bonds tying them to other nuclei. Nucleons at the surface have fewer such bonds, and the net result is that the total number of bonds per nucleon is much less for nuclei with fewer nucleons, since most nuclei in such nucleons are at the surface.

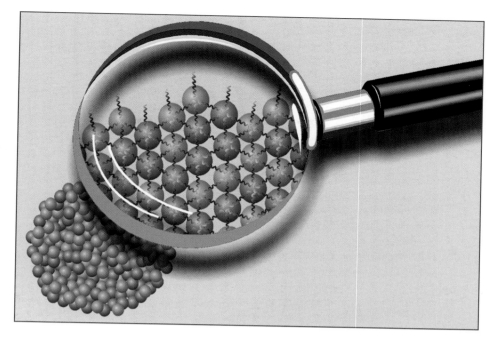

The short range of the nuclear force means that in any nucleus, the nucleons in the surface 'feel' the attractive force of a smaller number of other nucleons than those deep within a nucleus. In light nuclei, a higher fraction of the nucleons are near the surface so, overall, these nuclei have less binding energy than heavier ones.

One way of expressing this is by saying that light nuclei have extra energy, called surface energy, making them more weakly bound. All nuclei, visualized as droplets of nuclear matter, would have the same average energy per nucleon were it not for additional effects such as the surface energy, which is largest for the lightest nuclei, making them less stable. This is reflected in the steep upward rise of the floor of the valley at the low Z end.

The slow increase in energy, and decrease in stability, towards the heaviest nuclei is mainly due to the electrostatic repulsion between the protons. Although the electrical force is much weaker than the nuclear force, its longer range makes it important for larger nuclei. Every proton in a nucleus feels the attractive nuclear force only of its nearest neighbours, but feels the repulsive electrostatic force of every other proton in the nucleus. This repulsion therefore builds up as the number of protons increases, so that nuclei become less tightly bound as Z increases.

Moving up from the bottom of the valley floor, energy increases as the proton number and neutron number change for a fixed number of nucleons, in this case 137. The proton number is shown after the name of the element, but all elements are ^{137}Te, ^{137}I, etc.

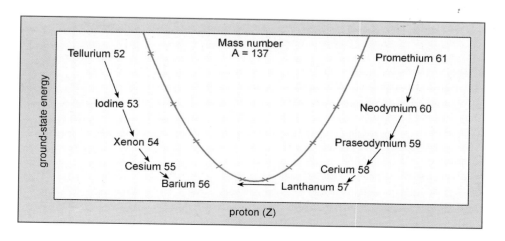

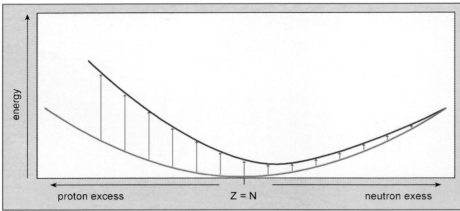

If there were no repulsion between protons, the energy would be lowest when the number of protons, Z, was the same as the number of neutrons, N. The most stable nuclei would have the same number of protons and neutrons as shown by the green line. This tends to be true for nuclei with less than 20 protons. Where there is a proton excess, there is additional energy, indicated by the red vertical lines. This is due to the repulsion between all the protons and is larger the greater the excess of protons. The total energy is shown by the blue line, and at its lowest point there are more neutrons than protons. This is why heavy nuclei have more neutrons than protons.

Electrostatic repulsion plays a role in other features of the energy valley. Consider the 'parabolic' down-and-up shape of the energy for a series of nuclei with the same total number of protons and neutrons. For a heavy nucleus, the minimum energy (most stable nucleus) occurs when there is a particular, considerable, excess of neutrons over protons. In quantum mechanics, if the protons did not repel each other then the minimum energy would occur for equal numbers of protons and neutrons. This energy would rapidly increase when there is an unequal number of protons and neutrons. The effect of the electrostatic repulsion between the protons is dramatic: the minimum energy is increased for all nuclei, but far more for nuclei with a large excess of protons. Thus the minimum energy occurs for nuclei with more neutrons than protons. These effects become larger for heavy nuclei, which have more mutually repelling protons.

The nature of the nuclear force and the quantum rules explains the overall shape of the energy valley: the way it arcs away from the diagonal N=Z line towards the neutron rich side, and the slow rise of the valley floor up towards the heavy nuclei.

More about nucleon orbits

The picture of the nucleus as a droplet of nuclear matter is incomplete. Such a picture cannot explain the gullies crossing the valley at the magic numbers. To understand these we require the completely different picture of a nucleus as a collection of nucleons in orbit.

A very familiar collection of bodies in orbit is the Solar System with planets orbiting the Sun. A nucleus is very different. Firstly, there is no heavy central object like the Sun holding the nucleons in orbit. The force each nucleon feels is an average force due to all the other nucleons. Secondly, of course, protons and neutrons exist in the tiny world where quantum rules are supreme. According to these rules, it is not possible to say that a proton or neutron is 'at' some particular place, something certainly possible with planets. All that can be said is how likely it would be to find a nucleon at a particular place at a particular moment. The nuclear equivalent to the planetary orbit is the quantum orbit described by a wave function specifying the places where a nucleon would be found if its position were measured.

The fact that no more than two protons or neutrons can occupy each orbit is crucial. The idea of orbits being 'occupied' is a key to the magic numbers. In the same way that electron orbits in a hydrogen atom only have particular energies, each proton orbit corresponds to a possible energy for a proton within the nucleus. The same is true for neutrons. This means there are families of nucleon orbits with similar energies with the number of orbits in each family fixed by quantum rules.

Period																		
1	1 H																	2 He
2	3 Li	4 Be											5 B	6 C	7 N	8 O	9 F	10 Ne
3	11 Na	12 Mg											13 Al	14 Si	15 P	16 S	17 Cl	18 Ar
4	19 K	20 Ca	21 Sc	22 Ti	23 V	24 Cr	25 Mn	26 Fe	27 Co	28 Ni	29 Cu	30 Zn	31 Ga	32 Ge	33 As	34 Se	35 Br	36 Kr
5	37 Rb	38 Sr	39 Y	40 Zr	41 Nb	42 Mo	43 Tc	44 Ru	45 Rh	46 Pd	47 Ag	48 Cd	49 In	50 Sn	51 Sb	52 Te	53 I	54 Xe
6	55 Cs	56 Ba	71 Lu	72 Hf	73 Ta	74 W	75 Re	76 Os	77 Ir	78 Pt	79 Au	80 Hg	81 Ti	82 Pb	83 Bi	84 Po	85 At	86 Rn
7	87 Fr	88 Ra	103 Lr	104 Rf	105 Db	106 Sg	107 Bh	108 Hs	109 Mt	110	111	112	113	114	115	116	117	118

57 La	58 Ce	59 Pr	60 Nd	61 Pm	62 Sm	63 Eu	64 Gd	65 Tb	66 Dy	67 Ho	68 Er	69 Tm	70 Yb	
89 Ac	90 Th	91 Pa	92 U	93 Np	94 Pu	95 Am	96 Cm	97 Bk	98 Cf	99 Es	100 Fm	101 Md	102 No	

1	2	3	4	5	6	7	8	9	10	11	12	13	14	15	16	17	18

Group

The chemical properties of the elements have a periodic behaviour. For example, the elements on the far right of the periodic table shown above – helium, neon, argon, krypton, xenon and radon – are all inert gases. They are particularly stable and do not need to form compounds with other elements. This is because their atoms have filled shells of electrons.

As protons are added to a nucleus, the orbits are filled family by family. When all the orbits in a family are full, the nucleus is especially stable and has a lower energy than it would have otherwise. The corresponding number of protons is a magic number, and the energy of nuclei with this number of protons will dip down below the average energy of nearby nuclei – such nuclei lie in a gully crossing the energy valley. For example, since 28 and 50 are consecutive magic numbers, the family of orbits that starts with the 29th proton, and is filled with the 50th proton, must have $50 - 28 = 22$ members. That is, it is a family of orbits that can accommodate 22 protons.

Neutron magic numbers arise in exactly the same way. The only difference is that for heavier nuclei the neutron orbits fill more easily since they do not have to worry about electrostatic repulsion. For example, the especially stable lead isotope ^{208}Pb has magic number 82 protons and magic number 126 neutrons.

Something very similar occurs with atoms. These are also particularly stable when all the places in a family of electron orbits are filled. These are the atoms of the inert or noble gases.

Nuclear shapes revisited

Liquid drops have a strong tendency to be spherical. This can be seen in film clips showing droplets floating in the gravity-free environment of an orbiting space station. For water droplets, as well as for nuclei, a particle in the surface feels the attractive force of fewer other particles than would be felt by a particle in the interior. So it is 'energetically favourable' to minimize the surface area, and spheres have the least possible surface area for a given volume. According to this argument, nuclei should also be spherical.

Many nuclei, however, are not spherical. This is a result of the nucleon orbits. The magic numbers mentioned above, 2, 8, 20, 28, 50, 82 and 126 apply to spherical nuclei. It turns out that there are other numbers that are 'semi-magic' if the nucleus has a non-spherical shape (referred to as deformed). This occurs because, for particular neutron numbers, the orbits for the last few neutrons are naturally deformed. For example, the isotope of the element of samarium with 86 neutrons, ^{148}Sm, is a typical spherical nucleus, but the isotope with 90 neutrons, ^{152}Sm, is deformed like an American football.

The energy of these orbits is lowered when the nucleus as a whole becomes deformed. As a result, nuclei with particular combinations of neutron number (or proton number) and American-football shape have lower energy than they would have otherwise. Such nuclei prefer not to be spherical. The effect of these semi-magic numbers on the energy valley is not as dramatic as the effect of the spherical magic numbers, but it is sufficient for many nuclei to be at their most stable when deformed.

For spherical nuclei, the magic numbers are well defined, so 30, 52 or 80 are definitely not magic although 28, 50 and 82 are. The number 90 marks the beginning of a whole range of neutron numbers for which the stable nuclei are deformed. This also happens with proton number 90 (thorium) so that thorium, uranium (Z=92) and plutonium (Z=94), the nuclei particularly associated with nuclear fission, are all deformed. This deformation, even for their most stable isotopes, gives these nuclei a start on the road to fission.

A deformed nucleus can be made to rotate, and this makes the phenomenon of super-deformation possible. When a nucleus rotates very fast, the resulting centrifugal force felt by its nucleons lowers the energy of particular highly non-spherical orbits. This results in the nucleus becoming stretched in a super-deformed shape.

Nuclear fission revisited

Today, the two pictures of nuclei – as droplets of nuclear matter, and as nucleons in orbit – are seen as complementary aspects of a more comprehensive picture. Both models provide useful insights in particular situations, for example, they both play a role in the description of nuclear fission.

The outermost proton and neutron orbits of a uranium nucleus have a lower energy when the nucleus is deformed. The energy would be higher if the nucleus were either spherical or more deformed. The energy for such different shapes can be calculated.

The energy is lowest when the nucleus has the slight American-football deformation that uranium nuclei are found to have. To stretch a uranium nucleus until it is more deformed than its natural shape requires a great deal of energy. This can be thought of as an energy 'hill'. Once over the hill, the energy in the nucleus falls rapidly. Thus if the nucleus could become deformed just enough to get to the top of the energy hill, it would then split apart into two fragments.

The nucleons in each fragment would no longer feel the attraction from the nucleons in the other fragment, but they would still feel the effect of the electric charge of the other fragment's protons. The two fragments would thus be thrust apart with great force. The fact that there is less energy within the two fragment nuclei than the original nucleus means that a great deal of energy has been released. This becomes the energy of motion of the two fragments which ends up as random motion of the surrounding atoms (or heat).

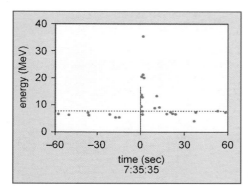

A pulse of neutrinos from supernova SN 1987A arrived on cue at the Kamiokande observatories underground neutrino detector in Japan. The designation 1987A identifies the supernova as the first to be observed in 1987.

The only problem with this picture of nuclear fission is that the nucleus does not have enough energy to reach the top of the energy hill. Quantum mechanics provides a way for this to happen by permitting a small probability of the nucleus tunnelling through the energy hill leading to spontaneous fission. This is the same tunnelling process by which an alpha particle emerges from a nucleus in alpha decay.

Sometimes energy from outside the nucleus makes it easier to surmount the fission barrier. This is what happens when a ^{235}U nucleus absorbs a neutron. The incoming neutron places the resulting 'compound nucleus' ^{236}U in an excited state and the extra energy allows it to overcome the fission barrier.

Neutron stars: the largest nuclei

One force of Nature has been ignored completely in our account of nuclei, the force of gravity. It is 10^{40} times weaker than the electrostatic force and makes absolutely no measurable contribution to the binding of atomic nuclei. It is certainly important to us, however, being the reason we do not require tethering to Earth with ropes.

Gravity differs from the electromagnetic force in that there are no such things as 'like' and 'unlike' gravitational charges: everything attracts everything else and there is no repulsion. In spite of the electromagnetic force being immensely more

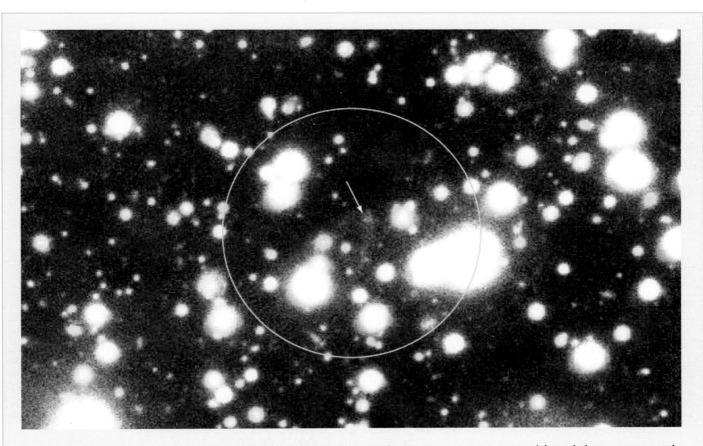

The arrow points to a neutron star at the head of the 'bowshock nebula', created as it sweeps at enormous speed through the tenuous gas and dust of space. It is rather like the bow wave of a ship travelling through the sea. The star itself is the very faint, blue object near the very tip of the cone. Most neutron stars are not visible, but this one is at a temperature of about 700,000 degrees, and it is probably heated by the very high speed impact of gas and dust. This is the closest known neutron star to the Sun, and presents a wonderful opportunity to study neutron stars. (Courtesy European Southern Observatories.)

powerful than gravity, almost everything around us on the macroscopic scale is neutral and electric forces mostly go unnoticed. When natural processes do separate some positive and negative charge, they tend to get back together quite dramatically. An example of this is lightning.

In spite of the weakness of gravity, there do exist 'nuclear' systems in the cosmos where gravity plays a crucial role. These are neutron stars. When a giant star, many times the mass of the Sun, comes to the end of its life, it undergoes a colossal explosion, becoming a supernova and briefly shines as an entire galaxy. If the conditions are right, it leaves behind a compact object about 10 to 15 km across (the size of a city) but with a mass at least 1.4 times that of the Sun. It consists largely of neutrons, the protons in the core of the star having combined with electrons under the pressure of the explosion to form neutrons and neutrinos. The neutrons remain as a neutron star. Vast quantities of neutrinos spread through space at the speed of light. In spite of their weak interactions with matter, a few might be detected on Earth in underground detectors like Kamiokande in Japan.

Surprisingly, perhaps, many of the properties of neutron stars can be understood directly from the properties of nuclear matter within nuclei. This is a huge extrapolation since the largest atomic nuclei have about 300 nucleons. Gravity acting within neutron stars makes it possible to have a bound object with about 10^{53} times as many nucleons, most of which are neutrons. It is remarkable that information from the shape of the energy valley can be extended all the way to neutron stars.

A crucial property is the incompressibility of nuclear matter. The pressure due to gravity acting at the centre of an object 10 km across, but with a mass greater than the Sun, is enormous. In spite of this pressure, the average density of a neutron star is about the same as that of an atomic nucleus: about 3×10^{14} times as dense as water. Only at the very centre does the pressure make it perhaps two or three times as dense as ordinary nuclei.

While the general properties of neutron stars are reasonably well understood, they raise many interesting questions that have yet to be answered. The full understanding of neutron stars demands a more detailed knowledge of the behaviour of nuclear matter at high temperatures and pressures. This is a powerful motivation for many experiments at accelerator laboratories such as RHIC, GSI and elsewhere in which high energy heavy nuclei are smashed together, producing for a brief instant some of the conditions within both neutron stars and the explosions which gave birth to them.

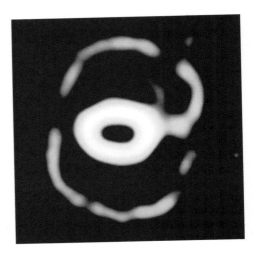

Shells of star material ejected into space from the star that left behind supernova 1987A in the Large Magellanic Cloud. The neutron star at the centre has not yet been identified. (Courtesy European Southern Observatories.)

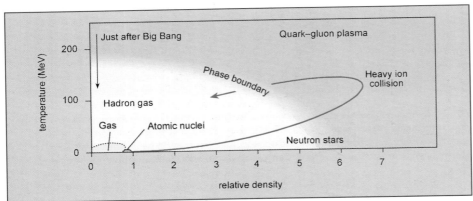

All nuclei have much the the same density, and unless they have been heated up in a nuclear collision they are 'cold'. But nuclear material can be compressed to a higher degree in neutron stars and heated up in supernova explosions. The Big Bang was exceptionally hot, and nuclear material then took the form of quark–gluon plasma, which is a different phase of nuclear material; much as ice and steam are different phases of water. In a collision between heavy nuclei at very high energies, nuclear matter will follow the path of the green line. This takes it out into the quark–gluon plasma region.

Jocelyn Bell Burnell and the Cambridge radio telescope. (Courtesy Jocelyn Bell Burnell.)

In the late 1960s quasars (quasi-stellar radio sources) had recently been discovered to be very distant, luminous objects which allowed the study of an early phase of the Universe. A radio survey of the sky, to discover more of them was started in Cambridge (UK). The radio telescope specially built, in-house, for the project spread over 2 hectares and consisted of 2048 dipoles operating at 81.5 MHz. The signals received by the radio telescope were recorded on strips of paper chart (no computers were available). The sky was repeatedly surveyed, each scan lasting 4 days, and producing 120 metres of paper chart. As the research student on the project I was responsible for running the survey and analysing the paper charts.

Occasionally on those 120 metres of chart there was about 0.5 centimetres of signal that puzzled me. It did not look like a quasar, nor did it did it look like locally generated radio interference that intermittently swamped the weak cosmic radio signals. The curious signal was very weak, was often not present, and moved across the sky with the stars.

Just as I focussed attention on it, it weakened and dropped to a level where we could not detect it. For a month I made special but fruitless visits to the observatory. Then it reappeared, and I could see that the signal was a string of pulses, equally spaced at about 1.3 seconds. The next month was one of worry; this signal was so unexpected we suspected something was wrong. However, it was not a fault with the radio telescope or receiver, for a colleague's telescope could pick it up too. Was it a satellite in an unusual orbit, or radar signals bouncing off the Moon and into the telescope? No. The pulse repetition rate was fast, implying a small object, but the pulse rate was very accurately maintained, implying a big object with large reservoirs of energy. Then we got an estimate of the distance to the source, which placed it far beyond the Solar System, but well within our Galaxy.

Then I discovered a second such source. It was in a different direction in space, and pulsed at a slightly different rate, but appeared to be of the same family of objects. It was immediately clear that we had stumbled over an unsuspected type of star. A few weeks later I discovered a third and a fourth.

Over the next six months we realized that these pulsars (pulsating radio stars) were neutron stars, sweeping a radio beam around the sky each time they rotated. Neutron stars were both small (in diameter) and big (in mass). Now over 1000 are known.

9. Cosmic furnaces

The stars and the birth of elements

Life on Earth depends entirely upon the Sun. So vital is its light and warmth that, from ancient times, it has often been worshipped. More recently, we have come to understand that not only do we rely on our local star to sustain us, but that we actually exist because of other, long dead, stars. The Earth, and almost everything on it, is made from the ashes of giant stars that lived their lives and then exploded billions of years ago. The Solar System itself, to which the Sun, planets and humans belong, came into being around 4.5 billion years ago, condensing out of the gas and dust ejected by an earlier generation of stars in their death throes. There is an abundance of carbon, oxygen and iron, all vital to life, because carbon, oxygen and iron nuclei were produced and expelled as those early stars died.

How stars shine and how the elements were produced, are two stories linked by nuclear physics experiments. Nuclear processes in stars can be understood by studying them on Earth; we can even attempt to tap the same source of energy which powers the Sun.

Apart from the fact that we depend on the Sun, the points made above are far from obvious. For example, in the 1830s the philosopher Auguste Comte claimed that the composition of the stars was an example of something that would always be beyond human knowledge. This statement was quite acceptable before the birth of spectroscopy.

More than two millennia before Comte, Aristotle made a proclamation which soon became dogma: "Everything in the heavens," he said, "is made of perfect, unchanging and incorruptible 'quintessence'." One reason Galileo Galilei was unpopular with the authorities was that some of his discoveries challenged this view. The idea that experiments on Earth might mimic processes in stars contradicted what people had believed for two thousand years. Such experiments are now carried out; stars are not made of 'quintessence' – essentially, they are composed of the same basic material as us.

A great physicist of the Victorian era, Lord Kelvin, proclaimed that the Sun could not be more than 100 million years old – a correct deduction based on the sources of energy known at that time. Geologists and biologists had strong reasons to believe that the Earth was much older, and in 1903 Rutherford pointed out the apparently continuous energy flowing out of radioactive nuclei showed "the maintenance of solar energy no longer presented a fundamental problem."

Finally, the nuclear physics pioneer, George Gamow, believed all the elements were produced in the Big Bang. This was a sensible deduction based on scientific knowledge of the 1940s, but Nature is an inexhaustible source of surprises and one of these was the discovery of technetium in certain red giant stars. This short-lived element could not have been made in the Big Bang so some elements, at least, must be produced in stars. There is now a great deal of evidence that this is so.

When George Gamow showed how alpha particles tunnel out of nuclei with the help of quantum theory, he had at the same time revealed how pairs of nuclei could tunnel inwards and fuse together. This opened the door to understanding how energy is produced in stars and how heavy elements are made from lighter elements, ultimately from hydrogen. He was also the father of Big Bang cosmology and the author of a much-loved series of popular physics books, still in print. (Courtesy Cambridge University Press.)

The Orion Nebula is found just below the belt in the constellation of Orion. It may be seen as a hazy patch in binoculars, or even with the naked eye from dark locations. Within it is a region of a thousand young stars crowded into a space less than the distance from our Sun to its nearest neighbouring stars. This is a stellar nursery where a new generation of stars is being born. This false colour image, taken in the infrared region of the electromagnetic spectrum, allows us to look deep into the region. (Courtesy European Southern Observatories.)

The abundance of the elements

Gamow was not entirely wrong. At present, the Universe is roughly 70% hydrogen, 28% helium and only about 2% everything else. (These percentages are by mass, not by number of atoms.) The hydrogen and most of the helium, together with some lithium (element 3) were produced in the Big Bang, but every other element has been produced in stars. The 2% that Gamow was wrong about includes the carbon, oxygen, nitrogen, iron and all the other elements.

It is not a trivial job to determine the cosmic abundances of the elements, that is what proportion of each element, or what proportion of each kind of atomic nucleus exists in the Universe. Many places are untypical in that they will contain a higher concentration of certain species than elsewhere – for example a jeweller's shop would not be a suitable place to estimate the cosmic abundance of gold.

Meteorites and the spectra of stars are two important sources of information about the cosmic abundance of elements. Certain kinds of meteorites are known to date back to the formation of the Solar System, and these reveal a consistent pattern of element proportions (the Solar System abundances) which are in fair agreement with that found in typical stars. The composition of each star depends upon the star's history, but an overall pattern emerges.

10 mm

A messenger from space. The light-coloured 'chondrules' visible in this meteorite, Adrar 303, have elements in the same proportion as the gas and dust from which the Solar System condensed. The proportion of different nuclei is remarkably close to what is found in many stars, representing the cosmic nuclear abundance. (Courtesy Ian Franchi, The Open University.)

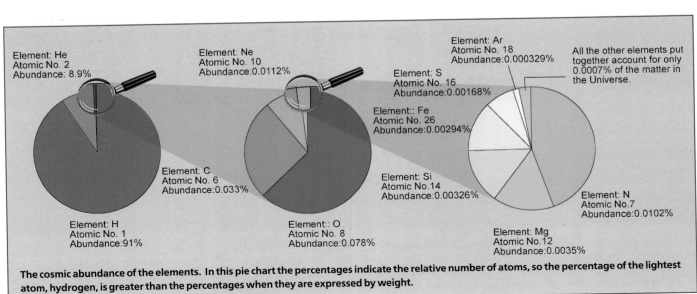

Element: He
Atomic No. 2
Abundance: 8.9%

Element: Ne
Atomic No. 10
Abundance:0.0112%

Element: Ar
Atomic No. 18
Abundance:0.000329%

All the other elements put together account for only 0.0007% of the matter in the Universe.

Element: S
Atomic No. 16
Abundance:0.00168%

Element:: Fe
Atomic No. 26
Abundance:0.00294%

Element: C
Atomic No. 6
Abundance:0.033%

Element: Si
Atomic No.14
Abundance:0.00326%

Element: H
Atomic No. 1
Abundance:91%

Element:: O
Atomic No. 8
Abundance:0.078%

Element: Mg
Atomic No.12
Abundance: 0.0035%

Element: N
Atomic No.7
Abundance:0.0102%

The cosmic abundance of the elements. In this pie chart the percentages indicate the relative number of atoms, so the percentage of the lightest atom, hydrogen, is greater than the percentages when they are expressed by weight.

There are vast differences in the proportions of each element. Clearly there is some influence from the energy valley, with elements located beyond iron and nickel being much less abundant than the lighter elements. Magic number elements tend to be more common.

By and large, the lightest elements are the most common. Somehow, the elements must come from a series of nuclear processes that start with hydrogen and build progressively heavier nuclei. There is a lot of carbon and oxygen because, in many stars, the sequence of nucleus building can only reach as far as these elements.

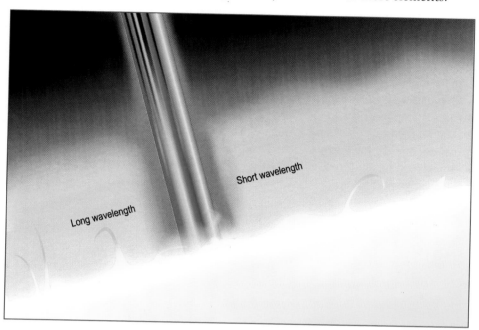

As light streams up from the hot surface of the Sun certain colours are absorbed by the atoms in the Sun's atmosphere. This process is symbolized by showing how progressively darker bands appear in the spectrum of white light as the light penetrates outwards through the Sun's atmosphere. When light is analysed on Earth, it shows the fully developed black absorption lines due to the entire thickness of solar atmosphere.

Long wavelength

Short wavelength

From hydrogen to helium

Almost all the energy we receive from the Sun comes from a series of reactions in which protons (hydrogen nuclei) interact to make helium nuclei. These reactions are also the first step towards the production of all the other elements. The process begins when two protons are attracted to each other and make a new nucleus, but there are two factors which work against this happening. The first is that although the nuclear force is strong and attractive when the protons are close, it is actually very hard for the protons to get near enough to each other for the force to be felt. This is because of the other force between protons, the electrostatic repulsion, keeping the two like-charged protons apart. The second problem is that the nuclear force is not quite strong enough to bind two protons; a nucleus consisting of two protons, ^{2}He, does not exist.

Neither of these problems exist if we begin with binding a proton and a neutron together. Since the neutron carries no electrical charge, there is no electrostatic repulsion acting against this pairing, and protons and neutrons do form a 'bound state': the deuteron (the nucleus of the isotope of hydrogen ^{2}H). While this would have happened during the Big Bang when there were free neutrons around, all those that did not pair up with protons will have undergone beta decay long ago. Free neutrons cannot survive longer than a few minutes before turning into protons. With none of these free neutrons remaining today, we are restricted to the fusion of two protons.

The electrostatic repulsion between two protons can be overcome in stars by a combination of quantum mechanics and high temperatures. The same rule of quantum mechanics which allows alpha particles to tunnel out of certain nuclei, enables protons with sufficient energy to tunnel through the force barrier between them. Protons at room temperature would certainly not have sufficient energy, but the temperature at the centre of the Sun is about 15 million degrees Kelvin. Consequently, a few protons do have enough energy to surmount the electrostatic barrier.

Having surmounted the barrier, the second problem of two protons not sticking together to form a nucleus is overcome by one of the protons turning into a neutron via beta decay, emitting a positron and a neutrino.

This complete process requires three of the four forces of Nature acting in coordination against the fourth. The gravitational force first compresses the hydrogen gas, raising its temperature and giving the protons the necessary energy to tunnel together. Next, the strong nuclear force keeps them close together while the weak nuclear force causes one of them to turn into a neutron, allowing the formation of the deuteron. During the process, the electromagnetic force acts to keep the two protons apart.

On average, it takes billions of years for protons in the Sun to join up to become deuterons because the transformation of a proton into a neutron happens very slowly. Once the deuteron is formed, the next step is much more rapid. Within seconds of being produced, a deuteron and another proton will tunnel together through the barrier between them to form the isotope of helium, ^{3}He.

The ^{3}He nuclei last somewhat longer than the deuterons, but they too will quickly disappear. Being positively charged, two ^{3}He nuclei will repel one another, but if they have enough energy they have a chance of fusing. If they do this, not all six nucleons will stick together, but a ^{4}He nucleus (alpha particle) will be formed and the two redundant protons will be emitted. The overall effect of this series of reactions is for six protons to form an alpha particle, with two protons left over, a process known as the 'pp chain'.

The first step, the fusion of two protons, is the bottleneck of the pp chain. It takes place too slowly for it ever to be observed on Earth, but the reaction between two ^{3}He nuclei can be measured. These very difficult measurements, first made in a tunnel deep beneath the Gran Sasso Mountains in Italy, help provide a firm basis for the detailed models of the internal workings of the Sun and other stars which are such a major part of modern astrophysics.

For most of its life, the Sun is, in effect, a factory for turning hydrogen into helium. Eventually, this process must end as the hydrogen at the centre of the Sun, where it is hot enough for fusion to take place, becomes exhausted. The core starts to contract and becomes even hotter, reaching the region of 100 million degrees. This is hot enough for the common isotope of carbon, ^{12}C, to be produced via the 'triple-alpha process'. While this is happening, the Sun's surface expands and the Sun becomes a red giant, eventually ejecting some of its outer layers into space. Some of this material will find its way into the next generation of stars and possible planetary systems.

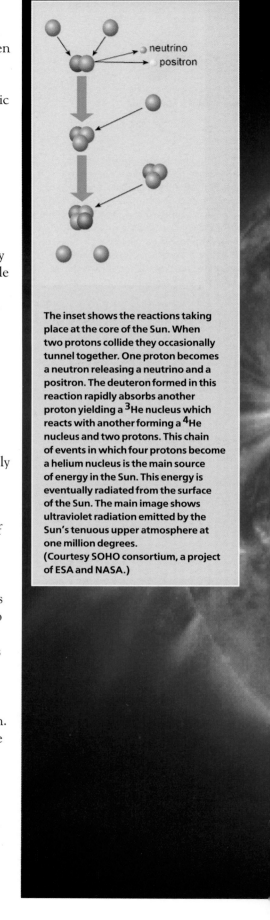

The inset shows the reactions taking place at the core of the Sun. When two protons collide they occasionally tunnel together. One proton becomes a neutron releasing a neutrino and a positron. The deuteron formed in this reaction rapidly absorbs another proton yielding a ^{3}He nucleus which reacts with another forming a ^{4}He nucleus and two protons. This chain of events in which four protons become a helium nucleus is the main source of energy in the Sun. This energy is eventually radiated from the surface of the Sun. The main image shows ultraviolet radiation emitted by the Sun's tenuous upper atmosphere at one million degrees.
(Courtesy SOHO consortium, a project of ESA and NASA.)

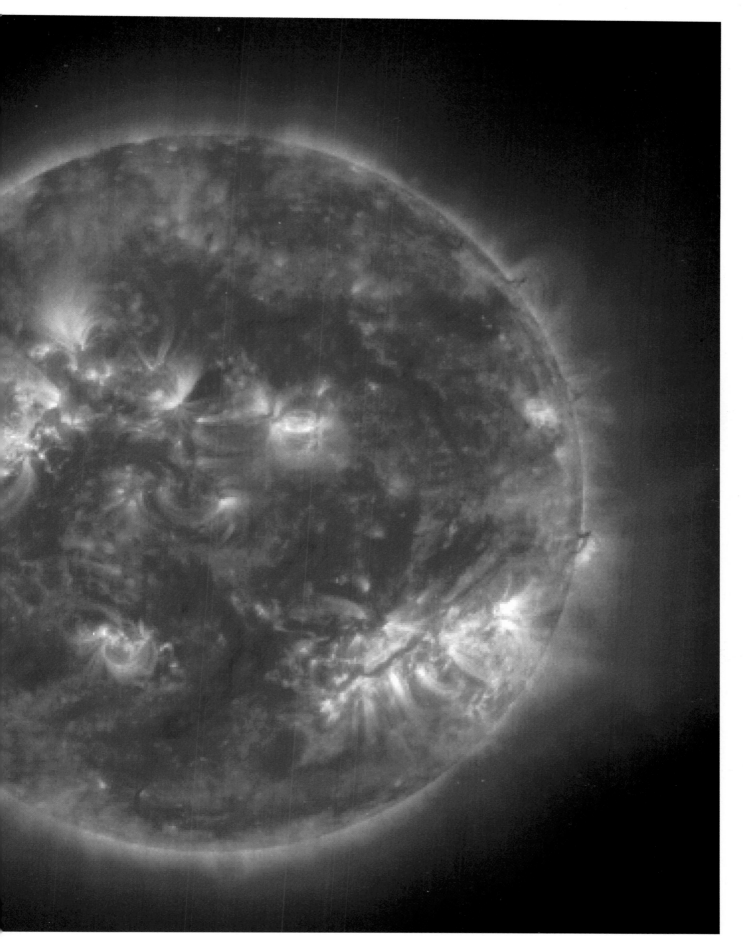

The Crab Nebula, M1, in the constellation of Taurus is the remnant of a supernova explosion at a distance of about 6000 light-years. First observed almost 1000 years ago, such explosions are the source of the heavy elements which we find on Earth. The green light is predominantly hydrogen emission from material ejected by the exploding star. The blue light is emitted by very high energy (relativistic) electrons spiralling in the large-scale magnetic field (synchrotron emission). (Courtesy European Southern Observatories.)

The triple-alpha process – a very delicate balance.

Like all carbon-based creatures, we depend on the production of carbon in stars. The natural way for ^{12}C to be produced would be for two ^{4}He nuclei to tunnel through the repulsive barrier between them to form a nucleus of the beryllium isotope ^{8}Be. Another ^{4}He nucleus would then tunnel in to make a nucleus of ^{12}C. Unfortunately, the ^{8}Be nucleus is not quite bound, and almost as soon as it forms it splits back into two alphas. It has the unimaginably small lifetime of about 10^{-16} seconds which gives very little opportunity for another alpha particle to join up with it to form carbon.

The situation is saved by 'resonance'. A resonance allows a radio to be tuned so that it is millions of times more sensitive to radio waves of a certain frequency than signals of all other frequencies. Similarly, nuclei have certain energies at which they are very much more likely to be formed in a reaction. ^{12}C has a resonance at exactly the right energy to allow alpha particles to exploit the extremely short-lived ^{8}Be nuclei in stars for carbon to be produced. Without this resonance at just the right energy, carbon-based life would not exist. It has been estimated that if the strength of the nuclear force were to be changed by one part in 200, then the production of ^{12}C nuclei would be thirty times less. The Universe is extraordinarily finely tuned!

Elements and energy in more massive stars

The Sun, along with other stars of the same mass or less, will never produce substantial amounts of nuclei heavier than ^{12}C, but such nuclei must have existed in the cloud from which the Solar System condensed.

The Sun is a star of about average mass. The stars that are much more massive produce the heavy elements. The more massive the star, the hotter it is and the more rapidly it passes through all the stages of its existence. The heaviest stars have quite short lifetimes and often end their lives in catastrophic supernova explosions. Such explosions are the source of much of the oxygen in our bodies and all of the uranium in our rocks and seas.

Temperature is the key to making heavy elements. All nuclei, being positively charged, repel each other. The greater the charge, the stronger this repulsive energy between them. The rate at which quantum mechanics allows them to tunnel through this energy barrier is very sensitive to how much energy they have due to their motion (kinetic energy) compared to the height of the barrier. A higher barrier means the nuclei need more energy to get through at a given rate. The energy of a nucleus within a star increases with the temperature of its location. The hottest part of a star is its core, at the centre. Since heat travels from hotter to cooler regions, the energy flows out from the core to the cooler outer regions of the star from where it is radiated into space.

At present the centre of the Sun is hot enough for hydrogen to fuse to make helium, but not hot enough for the triple-alpha process, and certainly not hot enough for the fusion of heavier nuclei. The temperature of the core will rise dramatically when it collapses after its hydrogen runs out. The triple-alpha fusion reaction ignites at about 100 million degrees. For stars more massive than the Sun, changes of this kind increase the core temperature even more, allowing much heavier nuclei to form.

Using carbon to make helium

In stars more massive than the Sun, there is an alternative to the pp chain method of producing helium from hydrogen. This is the CNO (carbon–nitrogen–oxygen) cycle, first proposed by Nobel prize winner Hans Bethe in the late 1930s. This process requires a star to have at least a trace of ^{12}C mixed in with the hydrogen. In this event, a ^{12}C nucleus can absorb a proton to produce a nucleus of the nitrogen isotope ^{13}N. Within a few minutes, the ^{13}N nucleus undergoes beta decay, turning into the carbon isotope ^{13}C, which is stable. Another proton will then tunnel through the ^{13}C barrier to produce ^{14}N. This nucleus is also stable and will eventually absorb another proton to become ^{15}O, a radioactive nucleus lasting just a couple of minutes before beta decaying to ^{15}N. The cycle finishes when ^{15}N absorbs a proton and an alpha particle is emitted leaving a ^{12}C nucleus, the starting point of the cycle.

The net effect of the CNO cycle is the same as the pp chain. Four protons, ordinary hydrogen nuclei, combine to produce a ^{4}He nucleus plus two positrons, two neutrinos and energy. In the CNO cycle, ^{12}C acts as a catalyst: it enables the process to take place, but is still there at the end. The CNO cycle certainly powers many of the stars shining today, but it would not have taken place in the very earliest stars that condensed out of the material produced in the Big Bang as that matter contained no carbon.

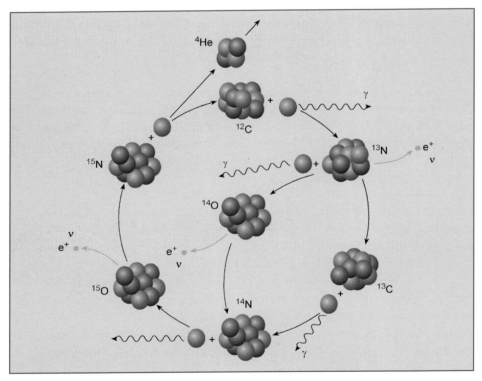

The CNO cycle takes place in hot stars. Four protons become an alpha particle (^{4}He nucleus) plus two electrons and two anti-neutrinos, indicated by the Greek letter nu (υ). Energy is released, partly as gamma rays (γ). The CNO cycle requires the presence of carbon nuclei ^{12}C, and the cycle starts when these absorb protons.

Using up the hydrogen

Towards the end of their lives, as the hydrogen in their cores becomes depleted, stars expand and become red giants. The Sun will enter its red giant phase in about five billion years and its outer atmosphere will extend to completely envelop the inner planets. Stars which are more massive than the Sun go through their red giant phase sooner in their lives, achieving particularly high temperatures deep within them as they do so. This enables a range of element-producing reactions to occur. Material ejected from the outer layers of red giant stars is believed to be the main source of the carbon and nitrogen upon which life depends.

Once ^{4}He and ^{12}C nuclei are created, a common process occurs by which they fuse to make ^{16}O; this requires a temperature of about 500 million degrees. At higher temperatures still, tunnelling between nuclei becomes easier and so a great many different reactions take place. For example, pairs of ^{12}C nuclei or ^{16}O nuclei can fuse together. Although there are many other reactions taking place, including those involving the absorption of protons, it is the occurrence of these processes which partly explains why nuclei which are multiples of the alpha particle, such as those of carbon and oxygen, are abundant on Earth.

Another reason for this abundance is that these nuclei are particularly stable. At temperatures around a billion degrees, there exists a huge flux of gamma ray photons that can break up nuclei, especially the less stable ones.

At the centre of stars much heavier than the Sun, there eventually comes a rather brief period when temperatures reach around 4 billion degrees. At this temperature ^{4}He can fuse with the isotope of silicon, ^{28}Si, and heavier nuclei. Nuclei of elements all the way up to iron are thus both created and destroyed, until there is an equilibrium and the most stable nuclei are synthesized. Some of these will be ejected into space. The shape of the energy valley means that very few of the elements lying much past iron and nickel are made in this process.

The lead in your car battery

There is no problem with producing elements with up to 26 or 28 protons: iron and nickel. Beyond these, all the way up to lead (with 82 protons), ways have to be found to make them.

Some red giants generate large amounts of neutrons. When a neutron strikes a nucleus it may go straight through, but often it will be absorbed. If the new nucleus is stable, it will simply shake off any excess energy as gamma rays and wait for the next neutron. On other occasions, the extra neutron places the nucleus too far up the neutron-rich side of the energy valley; the nucleus now has too many

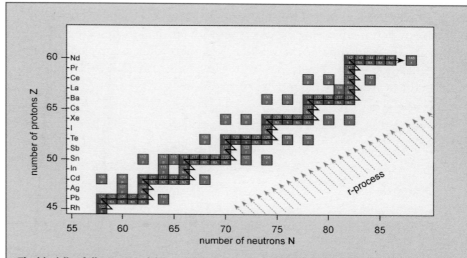

The black line follows part of the s-process from iron to bismuth, which takes about 100 years. When a nucleus absorbs a neutron the path steps one nucleus to the right. After a series of such steps, the nucleus undergoes beta decay, stepping diagonally up and to the left and then continues to absorb neutrons. The s-process occurs in red giant stars and the elements produced are indicated by blue squares. Some elements shown are made by other processes – red squares indicate elements made in the r-process (shown later in this chapter) in supernova explosions. Squares which are both red and blue indicate elements formed in both processes. Green squares represent nuclei made by the p-process, in which protons are captured.

neutrons, causing the nucleus to undergo beta decay. The neutron will become a proton and the Z of the nucleus will increase.

In these neutron-rich stars, the nucleus will continue to be bathed in neutrons and another neutron will be absorbed. The effect is that the nucleus works its way along the floor of the energy valley towards heavier and heavier nuclei. If the nucleus gets too high up the neutron-rich side, it will slide back down by emitting the electrons and anti-neutrinos of beta decay. This process, known as the 'slow process' or 's-process', produces many of the isotopes of elements heavier than iron and nickel.

The s-process cannot be the whole story. Not all isotopes of the heavy elements are made in the s-process. The gullies crossing the energy valley tend to stop the s-process happening. The gullies represent nuclei with certain numbers of neutrons which are particularly stable and have lower energies than their neighbours. Such nuclei with closed neutron shells do not readily pick up additional neutrons, so the path to new elements is halted.

The s-process can make lead, and is a major source of iodine, essential for life. Another element made in this way is short-lived technetium. The observation of the spectroscopic signature of technetium in stars is strong evidence for the s-process occurring within them since the half-life of technetium is a great deal shorter than the ages of the stars.

About half of the stable isotopes that are heavier than iron, including all the isotopes of uranium and thorium, cannot be made in the s-process. The existence on Earth of these elements is proof that there must be another process taking place. In addition, many elements have quite a few stable isotopes a short way up the neutron-rich side of the energy valley, and these are generally not made by the s-process.

When a star outshines a galaxy

Uranium, thorium and many other heavy nuclei are produced in the enormous explosions which are the dying throes of stars with ten or more times the mass of the Sun. These are supernovae in which, for a short while, the exploding star outshines a whole galaxy. Neutron stars are often formed in these explosions, while for the heaviest stars black holes are produced.

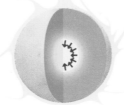

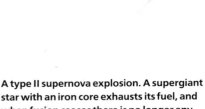

Supergiant

A type II supernova explosion. A supergiant star with an iron core exhausts its fuel, and when fusion ceases there is no longer any pressure to hold up the outer layers which collapse inwards. This causes huge pressure at its centre and the outer layers bounce out in a stupendous explosion releasing more energy at a stroke than the star has output in its entire life up to this point.

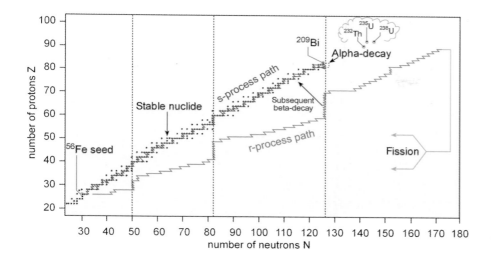

The huge flux of neutrons from a supernova leads to the r-process in which very short-lived nuclei are produced far up the neutron-rich side of the energy valley. These subsequently beta decay down towards the floor of the valley. All the uranium and thorium on Earth was produced by the r-process.

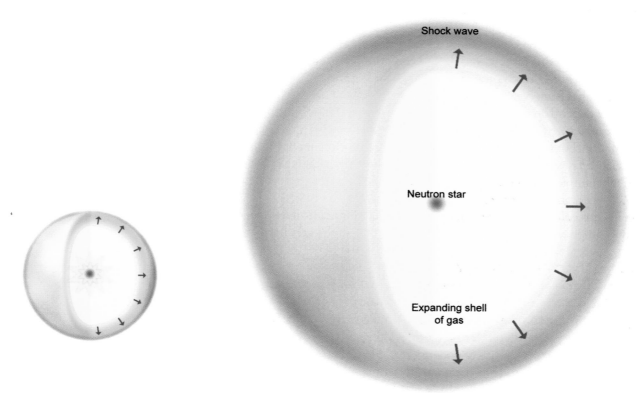

The high pressures cause electrons and protons to combine to form neutrons and an immense pulse of neutrinos producing a shock wave which ejects the outer layers into space, leaving behind a neutron star. If the original star is sufficiently massive, the result will be a black hole. The flux of neutrons leads to the production of new elements which are dispersed in space.

The element-making process associated with supernovae is known as the 'rapid' or 'r-process'. Certain (type II) supernovae release a vast flood of neutrons, far more than were present during the s-process. Exposed to such an intense flux, a nucleus can absorb a greater number of neutrons before it undergoes beta decay. As a result, the nucleus finds itself some distance up the neutron-rich side of the energy valley. Eventually, the nucleus becomes so unstable that it tumbles down the valley in a cascade of electrons and neutrinos. In this way, the r-process produces some nuclei that the s-process does not, uranium being an example.

Many nuclei on the r-process paths lie so far from the valley floor that they have never been made or studied on Earth. This has spurred intense activity to study nuclei out towards the neutron dripline. One goal is to get the nuclear information needed for a full understanding both of supernovae and the origin of the elements.

Fusion power on Earth

It has been a dream of scientists for half a century to reproduce and harness the fusion reactions that power the Sun. If successful, this process could provide a clean and inexhaustible supply of energy to help meet ever-expanding world demand.

The fusion of two hydrogen nuclei to produce helium is accompanied by the release of energy, but it is necessary to have enormous temperatures before the two hydrogen nuclei are energetic enough to fuse. Scientists face the problem of maintaining and containing such temperatures. In addition, the reaction employed cannot be that of two protons fusing together as it occurs far too slowly. Most practical schemes would have to involve the fusion of heavier isotopes of hydrogen such as deuterium and tritium. Fusion researchers use a machine called a Tokamak as a kind of test-bed for proving the principles of confined and controlled nuclear fusion on Earth, but while progress is being made, it will be many years before such an energy source is developed.

The big picture

We have now met most of the ways that nuclei are made in stars. The initial hydrogen from the Big Bang fuses within the hot cores of stars, producing helium, which adds to the helium already produced just after the Big Bang. When the hydrogen in a star has been consumed, a succession of changes begins. The details of these changes depend on the mass of the star. In the heaviest stars, quite large amounts of elements up to iron are produced. Beyond that there is a precipitous drop in production, but significant amounts of copper and zinc (elements 29 and 30) are produced. These elements are essential for life, but only a few beyond them are, such as selenium (34), molybdenum (42) and iodine (53).

For elements such as these, the stable isotopes with the largest neutron numbers tend to be made in the r-process, whereas the stable isotopes at the very bottom of the energy valley tend to be made by the s-process.

There are also other processes at work. For example, certain supernovae, known as type Ia, appear to produce huge fluxes of protons instead of neutrons. These give rise to the 'rapid proton' or 'rp-process'. This is similar to the r-process except that it operates on the proton-rich side of the energy valley. A nucleus absorbs several protons before it has a chance to undergo beta decay. The rp-process produces certain proton-rich isotopes that could not otherwise be made.

Another remarkable feature of these supernovae is the production of huge amounts of the unstable isotope of nickel, ^{56}Ni. This nucleus has 28 protons and 28 neutrons and is therefore a double closed shell nucleus. It is presumably formed by the successive fusion of 14 alpha particles, but a nucleus as far as ^{56}Ni from the floor of the energy valley cannot be stable even if it does have closed shells of both protons and neutrons. A ^{56}Ni nucleus has a considerable excess of energy, which it releases by beta decaying first to ^{56}Co (cobalt) and then to ^{56}Fe. When the star explodes, the energy released by these beta decays heats up the ejected material, contributing to the spectacular visible displays made by these supernovae.

Yet another process is responsible for producing certain light nuclei such as the lithium isotope ^{6}Li. Although a small amount of another lithium isotope, ^{7}Li, was made in the Big Bang, ^{6}Li was not. In addition, it is not made in stars – on the contrary, stars destroy it. The isotope ^{6}Li, along with the beryllium isotope ^{9}Be, are

JET, the Joint European Torus, is located at Culham, near Oxford in the UK. Deuterium and tritium ions, held in place by strong magnetic fields, are made to collide and fuse within the donut-shaped vacuum chamber. The cut-away shows the huge magnetic pole-pieces around the vacuum vessel. The height inside the vacuum vessel is more than 4 metres and the whole apparatus is 12 metres high. JET is just part of the long and difficult international effort to harness fusion energy for peaceful purposes on Earth.
(Photographs courtesy EFDA–JET.)

made by a process known as spallation, which follows the collision of a very high energy proton with a ^{12}C or ^{16}O nucleus in interstellar space. Such high energy protons are a major part of the cosmic rays that bombard the Earth from all directions. When the high energy proton strikes the heavier nucleus, it smashes it, producing a variety of lighter nuclei, known as spallation products, including ^{6}Li and ^{9}Be, which are not made elsewhere.

It is not surprising there is still uncertainty about the very complex processes in which heavy nuclei are made. Much of our understanding comes from painstaking measurements on unimaginably remote stars. A sample of material can never be taken directly from stars, but we do have access to matter produced in stars. After all, this is the matter of which we are composed.

The meteorites that sometimes fall to Earth are an even purer sample of the ash from exploding stars. Using data from meteorites and from the spectra of stars, astronomers and nuclear physicists have achieved an understanding of how the stars produce both heat and new elements. Many remarkable and surprising processes have been discovered and behind them all is the guiding hand of the nuclear energy valley.

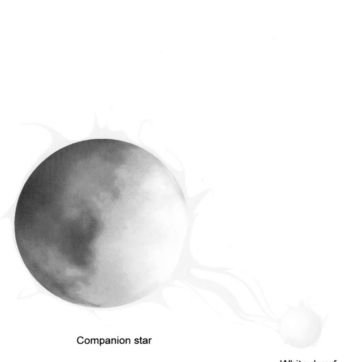

Companion star

White dwarf

A type Ia supernova explosion. Many stars come in pairs, so-called binary stars. Often, one of the stars is a compact white dwarf star. The maximum mass that a white dwarf can have is about 1.4 times the mass of the Sun. When matter is drawn to the white dwarf from its companion star and its mass is pushed over this limit, the star detonates in spectacular fashion.

The Crab Nebula is the remnant of a supernova that occurred in AD 1054 in the constellation of Taurus. At its heart is a pulsar. We now know that the pulsar is a neutron star, all that remains of the exploding star whose outer layers now fill vast regions of space.
(Courtesy European Southern Observatories.)

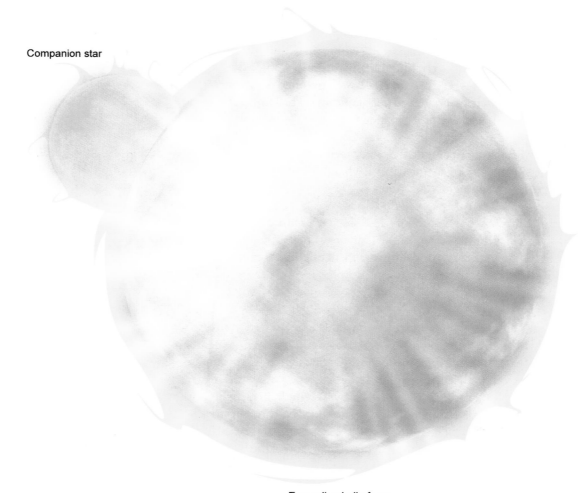

Companion star

Expanding ball of gas

10. Our violent origin

Cosmology and the nuclear processes in the Big Bang

Our Universe is getting larger. The space between the galaxies is expanding so that we see distant galaxies in all directions receding from us; the more distant a galaxy is, the faster it retreats. This is not because our galaxy occupies a privileged location at the centre of the Universe; an inhabitant of another, far away, galaxy would see the same spectacle.

If we imagine viewing the expansion in reverse, we can follow it all the way back to an instant, about 15 billion years ago, when all the matter in the Universe was squeezed together. This was the moment when space and time themselves came into being. What happened at that instant, and immediately after it, has become known as the Big Bang.

The Big Bang was an event like no other. Any attempt to describe it in terms of familiar concepts can only give a vague idea. For example, it is often referred to as an explosion of matter and energy, but this is quite wrong since the word 'explosion' suggests matter being forced out by huge pressures from a point in a pre-existing space. At the Big Bang, the whole of space itself was stretched out, carrying pure energy with it, as if it were being sucked out of nothing. The Universe did not expand into empty space; but space itself expanded.

Even stranger, the Big Bang did not take place at some moment in time, but instead marked the very beginning of time. To ask what there was before the Big Bang would seem to be a meaningless question.

General relativity is currently our best theory describing the nature of space and time. Its equations predict the expansion of the Universe, but it is not just mathematical equations that scientists rely on for their evidence. There are three independent pieces of evidence for the Big Bang.

Evidence for a Big Bang

Since all galaxies are composed of the same elements, the light emitted from distant galaxies should have the same characteristic wavelengths as that from nearer galaxies. The first clue that the Universe is expanding is that light from distant galaxies has a longer wavelength than it should. These longer wavelengths suggest the galaxies are receding from us at great speeds. The wavelength of the light that reaches us from those galaxies is 'stretched' as the space it travels through stretches. Longer wavelengths means light is shifted towards the red end of the visible spectrum. This stretching of space is just the ongoing expansion of the Universe that started with the Big Bang.

One way to visualize this expansion is to imagine blowing up a balloon with spots painted on it. No one spot would occupy the centre of the surface of the balloon, just as no one galaxy occupies the centre of the Universe. Concentrating on one spot and observing how other spots move in relation to it as the balloon is blown

Light from this distant cluster of galaxies, 1ES 0657-558, has taken about 6 billion years to reach us. When the light began its journey, the Universe was only about 60% of its current size. In other words, the distance between galaxies was only 60% of what it is now. Not only the Universe has stretched, the emitted light has too: wavelengths of light have increased by about 30%. Galaxies in this cluster appear 30% redder than they were when the light was emitted. (Courtesy European Southern Observatories.)

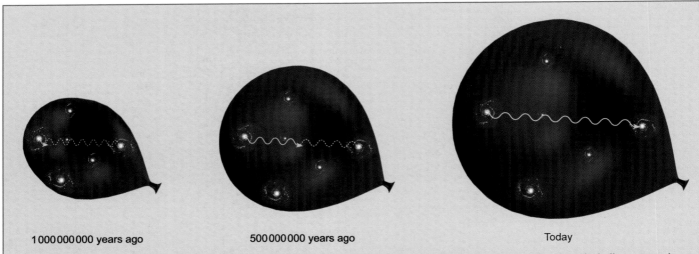

1 000 000 000 years ago   500 000 000 years ago   Today

The expansion of three-dimensional space may be visualized as the two-dimensional surface of a balloon being inflated. As the balloon expands, everything on its surface gets further away from everything else. The wavelength of light increases as the surface area increases. A balloon is not quite spherical, and has a place where the air is blown in. Space has nothing like that. Every location is like every other, and the Big Bang was everywhere.

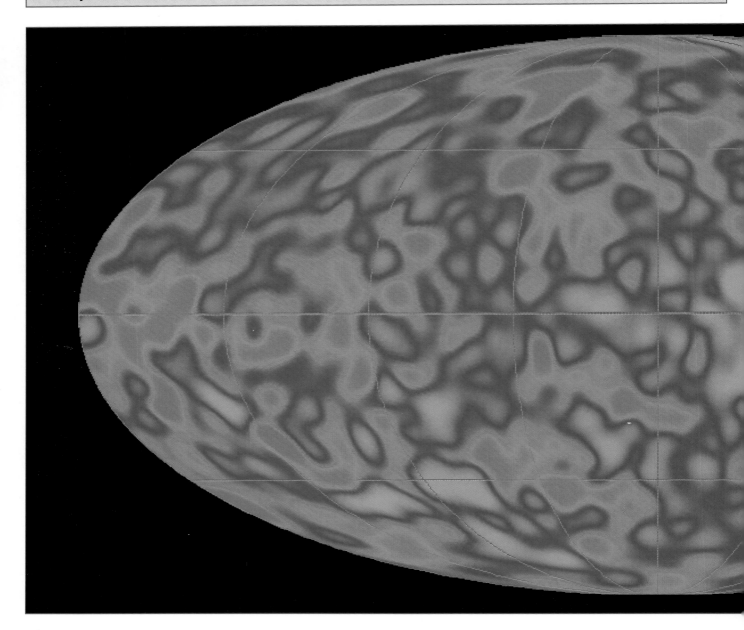

up shows every other spot recedes from it. The further away another spot, the faster it recedes, just like distant galaxies. Like all analogies , this one should not be taken too literally: the surface of the balloon is only two-dimensional.

The second piece of evidence for the Big Bang is the existence of microwave radiation approaching Earth from every direction in space. George Gamow first predicted the existence of this radiation long before it was actually observed. It is the 'afterglow' of the Big Bang. It has a temperature of less than three degrees above absolute zero (absolute zero is 273 degrees Celsius below freezing, about −459 degrees Fahrenheit), and consists of low energy photons. These photons existed about 300,000 years after the Big Bang, and remained when the Universe cooled sufficiently for them not to continue interacting strongly with matter.

The third, and just as important, line of evidence comes from nuclear physics. The relative proportions of hydrogen, helium and lithium fit very closely with the predictions of theoretical models of the Big Bang. However, the nuclei of these elements were not around at the very beginning. In fact, not even the nucleons themselves existed at the start.

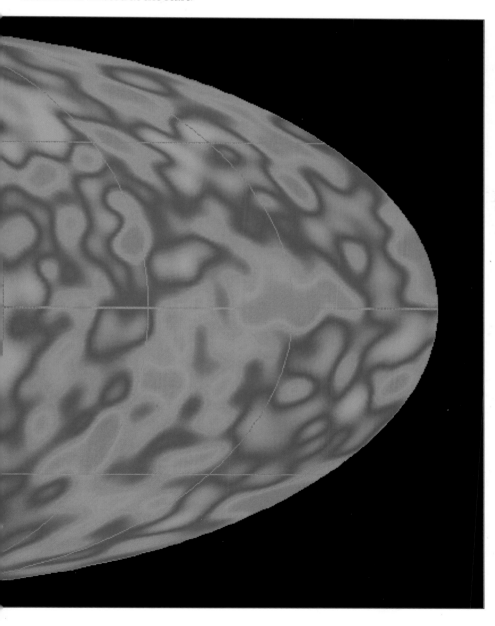

An image of the sky by COBE, NASA's Cosmic Background Explorer. The cosmic microwave background radiation predicted by Gamow is incredibly uniform, varying from one direction in the sky to another by only a few parts per million. The colours represent regions of the sky where the intensity is a little greater (red) or less (blue) than the average. When the radiation began its journey 300,000 years after the Big Bang it was not microwaves but visible light. Its wavelength has increased by a thousand times, just as the Universe has itself expanded a thousand times. (NASA)

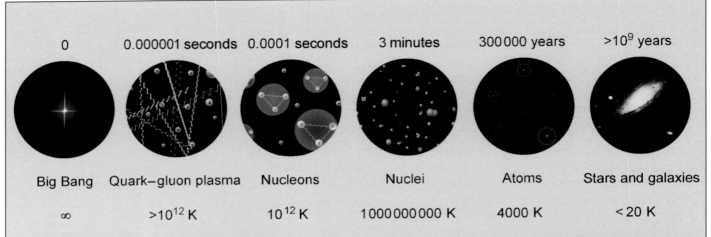

0	0.000001 seconds	0.0001 seconds	3 minutes	300 000 years	>10^9 years
Big Bang	Quark–gluon plasma	Nucleons	Nuclei	Atoms	Stars and galaxies
∞	>10^{12} K	10^{12} K	1 000 000 000 K	4000 K	< 20 K

Some highlights in the history of the Universe. The time along the top is the time after the instant of the Big Bang. The temperature along the bottom is the average temperature of the Universe as a whole.

Just after the beginning

The earliest meaningful time that can be comprehended with today's physics is the unimaginably short time of 10^{-43} seconds after the Universe came into being, when the temperature was over 10^{27} degrees. At this time it was expanding steadily. At 10^{-35} seconds, the Universe had grown to just 10^{-27} metres and the temperature had fallen to about 10^{27} degrees. At this moment it is thought that the expansion underwent a phenomenal acceleration. Exactly what happened is still the subject of debate among physicists, but it is believed that space itself rapidly inflated, carrying matter within it at a stupendous rate. During this 'inflation', the size of the presently visible Universe increased by a factor of at least 10^{25} from 10^{-27} metres to one centimetre (the factor may well have been as much as 10^{50} or even 10^{100}). This inflation had finished by 10^{-32} seconds.

Creating protons and neutrons

After the period of inflation, the Universe expanded at a much more sedate pace. By one millionth of a second after its birth, the temperature had fallen to below 10^{12} degrees and the nuclear ingredients, protons and neutrons, began to form.

Many properties of nucleons can be explained by assuming they are bundles of three quarks held together by gluons. The quarks come in six different types (flavours), only two of which are used in creating nucleons. Protons consist of two 'up' quarks, each of charge $+2/3$ and a 'down' quark with charge $-1/3$, giving an overall charge of $+1$. Neutrons have two down quarks and one up quark, making them neutral overall. The force-carrying gluons are electrically neutral.

There are strong reasons for believing that no single free quark or free gluon will ever appear in the laboratory. If enough energy is put into a proton to knock out one of its constituent quarks, instead of freeing it the energy creates a new particle that is itself made up of more than one quark. This is exactly what happens in collisions studied using particle accelerators.

The situation was different during the early Universe when all the matter was extremely compressed and hot. At one millionth of a second after the Universe came into existence, when the temperature was a million, million degrees, the boundaries between individual nucleons did not exist. Consequently, quarks and gluons could move freely in the quark–gluon plasma.

As the Universe continued its expansion, and the density and temperature fell, the quarks would have begun to arrange themselves into threesomes held together by gluons, and the nucleons were formed.

A quark–gluon plasma

The brief period before one millionth of a second into the Big Bang is not the only time that the quark–gluon plasma existed. In 2000, scientists at CERN in Switzerland published evidence, gathered from many years of experiments, strongly hinting that a quark–gluon plasma can be formed briefly when heavy nuclei collide head-on at very high energies, generating extremely high pressures and temperatures. These are just the conditions required for the boundaries between nucleons to dissolve, momentarily producing the quark–gluon plasma. This work is being continued at the Relativistic Heavy Ion Collider (RHIC) in the USA using even higher energies. It is intriguing to think that perhaps this unique state of matter has been produced just twice in the history of the Universe: first about a millionth of a second into the existence of the Universe, and then again briefly on planet Earth, 15 billion years later.

Making the light elements

Just after the Big Bang, although protons and neutrons attracted each other through the nuclear force, they did not immediately bind together to form deuterons. When the nucleons were first formed, the temperature was still too high and, for every nucleon there would have been several billion gamma ray photons that would have caused the disintegration of any deuterons as soon as they formed. Nuclei had to wait until the Universe was one second old before making even a fleeting appearance. At this time, the temperature had dropped to 10 billion degrees, but even at this temperature, deuterons are so flimsy that they are broken up almost immediately.

After about one minute, the Universe was cool enough for the deuterons to last long enough to have a reasonable chance of combining with a further proton or neutron to form a nucleus with three nucleons: either the hydrogen isotope ^{3}H (tritium), or the helium isotope ^{3}He. Both these light nuclei resist breaking up and can readily absorb a nucleon to form the much more stable ^{4}He nucleus.

The nuclear reactions leading to ^{4}He took place in a rapidly changing environment. The density was falling fast as the Universe expanded, so nuclei collided less frequently. The rapid drop in temperature meant there was less energy for charged nuclei to tunnel through the barrier between them. Neutrons are uncharged and so have the advantage of not having to tunnel through the barrier caused by the electric repulsion of the nuclei, but working against them is the fact that they steadily undergo beta decay, turning them into protons.

These changes meant such nuclear fusion reactions ceased after a few minutes and no more new nuclei could be made. The primordial matter formed a few minutes after the first instant of the Big Bang had just a small number of isotopes in proportions that remained fixed for many millions of years. Only when stars began to form did other quite different element-making processes occur.

The proportions of the primordial isotopes can be predicted by theory. The overwhelming part (76% by mass) consisted of hydrogen (single protons) with the remainder (24% by mass, or about 9% by number of atoms) being helium (two protons and two neutrons). There were also tiny, but significant, traces of other isotopes: a few tens of parts per million consisted of deuterium, slightly less of the helium isotope ^{3}He, and just a tenth of a part per billion of the lithium isotope ^{7}Li. No stable nuclei with five nucleons exist in nature, and the only stable nucleus with 6 nucleons (^{6}Li) had to wait until much later to be formed.

These predictions depend upon the density both of matter and photons at the time the processes were taking place. Most of the predictions seem to fit quite well with modern observations. For instance, ^{7}Li can be detected at the surface of very old stars lying at the edge of our galaxy which have not been contaminated by the ashes of exploding stars, and the amount of deuterium can be measured by the absorption of light from very distant galaxies as it passes through clouds of primordial gas that have not yet condensed into stars.

The Big Bang elements leave no scope whatever for the formation of heavier nuclei necessary for the rich and complex world in which we live or for the evolution of life. This has been made possible by the billions of years of nuclear reactions occurring within stars. Consequently, today there is somewhat less hydrogen (70% by mass) than there was in the early Universe, while the amount of ^{4}He has increased to 28% due to stellar fusion. In addition, a vital 2% of heavier elements has appeared.

The Big Bang, nucleon density and modern cosmology

Many questions remain unanswered: will the Universe expand forever or will the expansion stop some day, to be followed by contraction ending with a Big Crunch? Perhaps, as recently suggested, the expansion of the Universe is actually accelerating. Cosmologists debate these questions intensely. One key factor is just how much matter exists in the Universe. Astronomers find this a very difficult question to answer, but the abundance of primordial isotopes provides a clue. The present density of nucleons in the Universe depends on the density of matter and radiation at the time the primordial elements were being formed, and nuclear physics provides a way of deducing this primordial density.

Nuclear processes that took place in the first few minutes, in particular the amount of deuterium that was produced, depended crucially on how densely concentrated the nucleons were. The deuterium in existence today is primordial – made in the first few minutes after the Big Bang. This is known because any new deuterium created by reactions within stars is immediately destroyed by further reactions. Thus, if the amount of deuterium in existence today is measured, we can learn about the conditions at the time the primordial elements were made, which in turn provides a clue as to how much total matter there is today.

It turns out that the amount of deuterium is not exactly what is expected, but very close. It is not known if this is a consequence of problems with the measurements, which are difficult, or if cosmology is less of an exact science than some cosmologists would like to think. Maybe there are still a few surprises to come from nuclear physics.

These are exciting times, and nuclear physicists, astronomers and cosmologists, working together, will no doubt find an answer in the coming years. In doing so, we shall achieve a firmer understanding of the history of the Universe and also of its future.

Three possible ways in which the Universe might end. The picture is symbolic since the Universe does not have an 'edge'.

Left: if the density of the Universe is greater than a critical value, then it will eventually stop expanding, and will contract under the force of gravity, ending in the 'Big Crunch'.

Centre: if the density is neither too little nor too great, then the expansion of the Universe will gradually slow down. This is called the 'Flat' Universe.

Right: if the density is less than the critical value, it will expand for ever. This is called the 'Open' Universe. (Julian Baum)

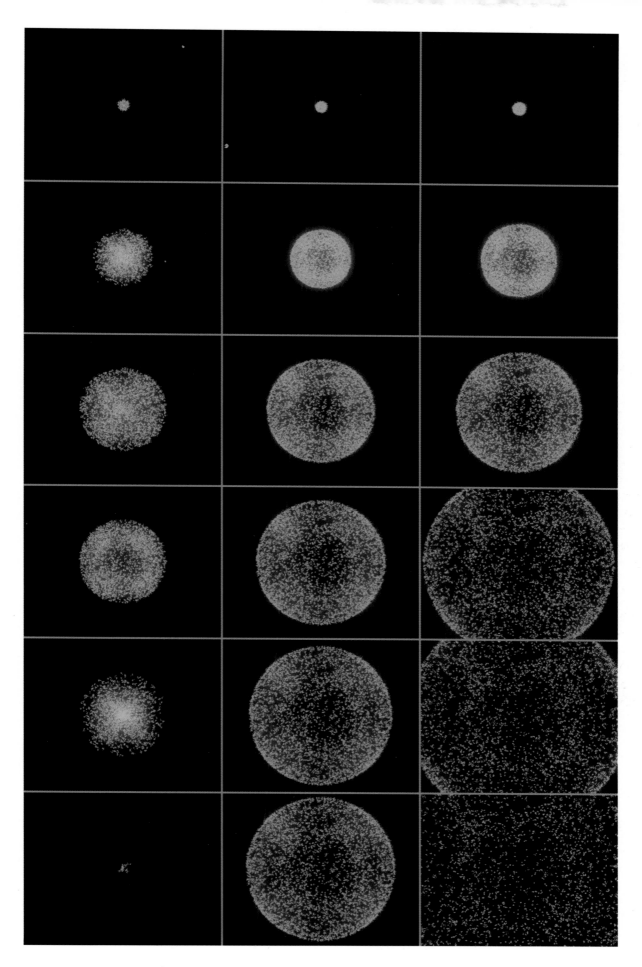

Further reading

The Magic Furnace: The Search for the Origins of Atoms, Marcus Chown (Oxford University Press, 2001)
A very readable and eloquent historical account of how scientists unravelled the mystery of atoms. It chronicles the achievements of the pioneers of atomic and nuclear physics.

The Quantum Universe, Tony Hey and Patrick Walters (Cambridge University Press, 1987)
A readable, lavishly illustrated book, in much the same spirit as ours. It is dedicated to the whole range of microscopic physics, so there is less on nuclear physics.

From X-rays to Quarks, Emilio Segrè (W.H. Freeman, 1976)
A first hand account of the history of modern nuclear and particle physics by the Nobel Prize winning student of Fermi. Most likely to be found in a library.

Enrico Fermi Physicist, Emilio Segrè (Chicago University Press, 1995)
A readable account of the life and works of Fermi by his distinguished student.

The Nature of Matter, J.H. Mulvey (ed.) (Clarendon Press, 1981)
A series of lectures by front rank physicists delivered at Wolfson College, Oxford. Most of the lectures are not too technical, and some are very inspiring.

Atom: Journey Across the Subatomic Cosmos, Isaac Asimov (Penguin, 1992)
The legendary Isaac Asimov starts with a simple query: How finely can a piece of matter be divided? This leads readers on a far-flung quest for a final answer, a search that encompasses light and electricity and their components - strange bits of matter that challenge our assumptions about the very nature of time and space.

The First Three Minutes: A Modern View of the Origin of the Universe, Steven Weinberg (Basic Books, 1993)
A classic of contemporary popular science writing by a Nobel Prize winning physicist explains what happened when the Universe began, and how we know.

Stardust, John Gribbin and Mary Gribbin (Penguin, 2001)
A popular account of nuclear astrophysics with diversions into astronomy, cosmology, geology, and even the possibility of extraterrestrial life.

In Search of Schrödinger's Cat: Quantum Physics and Reality, John Gribbin (Bantam, 1985)
While getting a little out of date, this book still provides one of the best and clearest explanations of quantum physics. An easy read for the layman with no background in the subject.

Nuclear Physics: Principles and Applications, John Lilley (John Wiley, 2001)
Although this is an undergraduate textbook, it is an excellent source for applications of nuclear physics, particularly in medicine.

Useful websites

Nuclear Physics Laboratories around the world

GSI (Darmstadt, Germany) http://www.gsi.de/

GANIL (Caen, France) http://www.ganil.fr/

CERN (Geneva, Switzerland) http://www.cern.ch/

RHIC (Brookhaven, USA) http://www.bnl.gov/rhic/

Jefferson Lab (Virginia, USA) http://www.jlab.org/

Argonne National Lab (Argonne, USA) http://www.anl.gov/

Lawrence Berkeley Lab (Berkeley, USA) http://www.lbl.gov/

TRIUMF (Vancouver, Canada): http://www.triumf.ca/

National Superconducting Cyclotron Laboratory (East Lansing, USA) http://www.nscl.msu.edu/

RIKEN (Saitama, Japan) http://www.rarf.riken.go.jp/rarf/rarf.html

Vivitron (Strasbourg, France) http://eballwww.in2p3.fr/vivitron/uk/welcome.html

CRC (Louvain la Neuve, Belgium) http://www.cyc.ucl.ac.be/

KVI (Groningen, Netherlands) http://www.kvi.nl/

ANU (Canberra, Australia): http://wwwrsphysse.anu.edu.au/nuclear/

JINR (Dubna, Russia): http://www.jinr.dubna.su/

INFN (Legnaro, Italy) http://www.lnl.infn.it/

TSL (Uppsala, Sweden) http://www.tsl.uu.se/

JYFL (Jyvaskyla, Finland) http://www.phys.jyu.fi/research/

ORNL (Oak Ridge, USA) http://www.ornl.gov

Nuclear Fusion Power

JET (Culham UK) http://www.jet.efd.org/

ITER (International) http://www.iter.org/

Physics Institutes and Societies

Institute of Physics (IoP) (UK): http://www.iop.org/

TIPTOP http://physicsweb.org/TIPTOP

American Physical Society (APS) (USA): http://www.aps.org/

Glossary

Italic type indicates terms with a separate glossary entry

Alpha particle: The nucleus of the helium atom containing two protons and two neutrons. It is a highly stable nucleus and is emitted in one piece from many heavier nuclei in a process known as alpha decay. Alpha particles were known before it was understood that they comprised of *protons* and *neutrons*. They are also doubly ionized helium atoms, that is helium atoms with both electrons removed.

Anti-matter: matter made up entirely of anti-particles.

Anti-particle: a type of subatomic particle that is like a mirror image of the original particle in that many of its key properties are reversed. For instance, the anti-proton has the same mass as the *proton* but is negatively charged. When particles and their corresponding anti-particles meet, they mutually annihilate each other in a burst of energy. Similarly, a particle and anti-particle pair can be created out of pure energy. The anti-electron is known as a *positron*, and has a positive charge with the same magnitude as the negative charge on an *electron*.

Atomic mass unit (amu): the conventional unit for expressing the mass of nuclei. It is one twelfth of the mass of a neutral carbon atom, ^{11}C. The mass of a nucleus in amu is approximately equal to the mass number of a nucleus.

Atomic number: symbol Z, the number of *protons* in a nucleus. Also the total number of *electrons* in a neutral atom, since if the charges are to balance, the number of electrons outside the nucleus must equal the number of protons in the nucleus.

Baryons: *hadrons* that are composed of three *quarks*. The lightest baryons are the *protons* and *neutrons*. Heavier hadrons, such as the *delta particle*, are unstable. Baryons and *mesons* (which contain just a quark and an anti-quark) make up the two types of hadron.

Beta decay: the process, governed by the weak nuclear force, whereby *protons* and *neutrons* can transform into each other. When a neutron beta decays, an *electron* and an anti-neutrino are released. A free neutron can beta decay because its mass exceeds the sum of the proton and electron masses. A proton can only beta decay within a nucleus with excess energy and becomes a *positron* plus a *neutrino*; this happens within nuclei with a sufficient excess of protons.

Beta ray: the old name given to the *electrons*, or *positrons*, emitted from nuclei in the process of *beta decay*.

Borromean nucleus: a certain type of unstable nucleus that acts as though it is made up of three distinct parts. The bulk of the nucleus makes up the core, while two *nucleons* (usually *neutrons*) 'float'

around outside it. These three constituents (core plus two nucleons) are held together very weakly by the strong nuclear force in such a way that if any one of them is removed then the force between the remaining two is too weak to hold them together and they too fall apart. This behaviour is unique in Nature. The term Borromean comes from the field of mathematics known as knot theory in which the Borromean rings are interlocked such that each one holds the other two together. Examples of Borromean nuclei are ^{6}He, ^{11}Li and ^{14}Be, all of which are also *halo nuclei*.

Bubble chamber: a device for revealing the paths of charged (highly *ionizing*) particles as they pass through a 'super-saturated' liquid. Sometimes a liquid, such as liquid hydrogen, can be heated above its boiling point if it is very pure. However, a charged particle passing through such a 'super-saturated' liquid will leave a series of bubbles at the points where the particle ionized the liquid.

Chain reaction: when a uranium nucleus is induced to undergo *fission* by the absorption of a neutron, it will itself release a few *neutrons*. These can then in turn induce other nuclei to undergo fission, and so on, throughout a significant volume of uranium. This is a neutron chain reaction.

Cloud chamber: a device for revealing the paths of charged (highly *ionizing*) particles as they pass through 'saturated' water vapour. There is a limit to the amount of water vapour air can hold, but that limit can sometimes be exceeded. A charged particle passing through such a 'saturated' vapour will leave a series of droplets at the points where the particle ionized the vapour.

Cyclotron: an accelerator for charged particles in which the particles are held in a spiral orbit in a vacuum chamber by a magnetic field, and given a series of pushes by an electric field.

Delta particle: a relation of the *nucleon*, this particle can be thought of as an *excited state* of the nucleon with a slightly greater mass.

Diffraction: the property of all waves whereby they spread out on encountering a obstacle. The amount of spreading depends on the *frequency* of the wave and the size of the object.

Driplines: these are the limits on a Segrè chart beyond which nuclei do not exist. So-called because it is as if nucleons inserted into such a nucleus drip straight out again.

Electromagnetic radiation: any radiation consisting of self-sustaining electric and magnetic fields. All electromagnetic radiation travels in a vacuum with the same speed: the speed of light. Light, radio waves, ultraviolet, gamma and infrared radiation are fundamentally the same, differing only in *frequency* and *wavelength*. All their differing effects on matter result from the difference in frequency and hence the difference in energy of the photons.

Electron: the first elementary particle to be discovered. It belongs to the class of particles known as *leptons*. Electrons are very light negatively charged particles and are the constituents of atoms outside the nucleus. Electrons have a mass of 9×10^{-31} kg, about one two-thousandth of the mass of the lightest atom, the hydrogen atom. They have zero size and so are considered to be 'point particles'. They are the carrier of electricity in metals.

Electron capture: the process in which an *electron* is absorbed by a nucleus wherein it combines with a *proton* to make a *neutron* and a *neutrino*.

Energy levels: the allowed discrete (quantized) values of energy that a nucleus can have. Each kind of nucleus has a unique pattern of energy levels. Atoms and molecules also have unique patterns of energy levels.

Energy valley (nuclear): not all nuclei have the same amount of energy per nucleon. If all nuclei are arranged according to N and Z, as in a *Segrè chart*, and a line is raised from the position of each nucleus, proportional to the energy per nucleon, then the tops of those lines will form a surface taking the appearance of a valley. The stable nuclei will be those near the bottom of the valley. The nuclei higher up the valley will undergo radioactive transformations, losing energy and sliding down the sides of the valley.

Excited state: any *energy level* of a nucleus above its *ground state*.

Fission: the process whereby a heavy nucleus splits into two roughly equal smaller nuclei, releasing energy locked up inside it. Usually, fission takes place after a nucleus has been stimulated by the absorption of a *neutron*, but *spontaneous fission* does exist. Fission is the process that produces nuclear power through controlled chain reactions.

Frequency: the number of vibrations of an oscillating system that occur in one second. For the case of waves, it is the number of wave crests that pass by a fixed point in one second. It is measured in Hertz (= cycles per second).

Fusion: the nuclear process whereby two light nuclei can overcome the mutual electric (Coulomb) repulsion to fuse together. This is accompanied by the release of a large amount of energy and is the source of energy in the Sun and other stars. It is hoped that fusion will one day be harnessed as an energy source on Earth.

Gamma ray: a high energy *photon* (particle of light) released from within atomic nuclei when they find themselves in an unstable excited state. Gamma rays can also be absorbed by nuclei which then become excited.

Gluon: massless particle, never seen isolated outside *hadrons*, which generate the attraction between the *quarks* within the hadrons, holding them together.

Ground state: the lowest *energy level* of a nucleus (or atom).

Hadrons: all particles that interact via the strong nuclear force. Hadrons are composed of *quarks*; *protons* and *neutrons* are hadrons, as are *mesons*.

Half life: the time after which half of a large collection of identical radioactive nuclei will have decayed.

Halo nucleus: a certain type of exotic nucleus discovered in the mid 1980s that has many more *neutrons* than the stable *isotope* of that element. This sometimes results in the outer one or two neutrons being very weakly linked to the rest of the nucleons and thus they spend much of their time far beyond the range of the strong nuclear force that originally bound them to the rest of the nucleus. Such nuclei are highly unstable and only exist due to strange rules of quantum mechanics. Examples of one neutron halo nuclei are ^{11}Be and ^{19}C, while two-neutron halo nuclei tend to be *Borromean*. Nuclei with *proton* halos also exist (such as ^{8}B) but in that case the repulsion between the proton's positive charge and the rest of the nucleus means that it cannot stray very far out or it falls out. Thus proton halos tend to be smaller than neutron halos.

Interference: a property of waves whereby two waves overlap each other to produce a pattern of maxima (when two wave crests meet and combine) and minima (when a wave crest meets a wave trough and they cancel out). The pattern of dark and light that results allows us to learn about the waves and about the system in which interference takes place.

Ion: an atom, or molecule, which is no longer electrically neutral, usually because one or more *electrons* have been knocked out, but the term can apply to an atom with an extra electron (negative ion).

Ionize: the process of removing *electrons* from atoms or molecules so that they are no longer electrically neutral. *Alpha*, *beta* and *gamma rays* ionize the atoms of matter with which they interact.

Isobar: *nuclides* with different numbers of *protons* and *neutrons* but the same total number of protons plus neutrons, i.e. the same atomic *mass number*.

Isomer: a nucleus that is in a long-lived excited state (called an isomeric state or a metastable state). Certain nuclei can remain in such excited states due to certain quantum properties they have which forbid them from dropping down to a lower *energy level* by emitting gamma rays.

Isotones: *nuclides* with the same number of *neutrons* but different numbers of *protons*.

Isotope: all *nuclides* of a particular element (having the same number of *protons*) but with different numbers of neutrons are known as isotopes of that element. Thus ^{12}C and ^{14}C are different isotopes of carbon.

Isotope shift: the optical spectrum of an atom depends almost entirely on the *electrons* outside the nucleus, but the size of the nucleus does have a tiny but measurable effect. This means that careful measurements of the optical spectra of atoms allow us to measure the size of the nucleus. This is very useful for nuclei that live for too short a time for measurements based on electron scattering to be made.

Leptons: one of the two classes of elementary particles in nature. They include the *electron*, muon and tau particle, along with their corresponding *neutrinos*.

Linear accelerator: an accelerator for charged particles in which the particles are given a series of pushes down a long straight vacuum chamber by oscillating electric fields.

Magic number: particular numbers of *protons* or *neutrons* lead to nuclei with increased stability compared to the neighbouring nuclei. For neutrons these numbers are 2, 8, 20, 28, 50, 82 and 126. They are the same for protons except that no nucleus with 126 protons is known.

Mass number: symbol A, the total number of *protons* and *neutrons* in a nucleus. A = N + Z, where N is the number of neutrons and Z is the number of protons.

Meson: subatomic particle that can be considered as the carrier of the strong nuclear force between *nucleons* in nuclei. There are various types of mesons and they can be neutral, positively or negatively charged. It is now known that mesons are, like nucleons, themselves composed of *quarks*. But unlike nucleons, which are made up of three *quarks*, mesons contain a quark and an anti-quark.

Neutrino: an effectively massless particle emitted during *beta decay*. Its name means 'little neutral one' and until recently was thought to have no mass at all (like the *photon*). It is now known that there are three types of neutrinos, along with their corresponding anti-particles, but it is the lightest of these that is emitted by nuclei.

Neutron: the other constituent, along with *protons*, of nuclei; it has a mass very slightly greater than the proton but is electrically neutral. Neutrons cannot survive for long outside the confines of the nucleus and they *beta decay*, that is turn into protons and anti-neutrinos, after about ten minutes.

Neutron star: the compact corpse left behind when a giant star dies in a *supernova* explosion. It has much the same density as a nucleus. Many experiments are aimed at understanding enough of the properties of *nuclear matter* at high pressures to enable us to understand neutron stars.

Nuclear matter: broadly speaking, the stuff of nuclei. It is because nuclear matter is incompressible that the density of *nucleons* at the centre of nuclei is much the same for all nuclei except the very lightest. For the same reason, neutron stars have much the same density as we find at the centre of nuclei.

Nucleon: general term for either a *proton* or a *neutron*.

Nuclide: a nucleus of a given number of *protons* and *neutrons*. There are about 7000 different possible nuclides, only several hundred of which are stable.

Oblate deformation: the deformation of a sphere that is achieved by squeezing two sides together. The Earth is slightly oblate since it is a little squashed at the poles making the equator slightly longer than it would be if the Earth were a perfect sphere.

Photomultiplier: a very sensitive *photon* detector, capable of measuring the energy in a very weak pulse of light. It forms part of a *scintillation detector* for measuring *gamma rays*. The pulse of light is generated as a *scintillation*.

Photon: a single particle of light. Einstein proposed in 1905 that light came in packets or light quanta, as an explanation for the photoelectric effect, whereby light can knock out electrons from the surface of a metal. Together with the work of Max Planck, this marked the beginning of the old quantum theory in which light is considered to be composed of discrete lumps. Later, it was realized that these packets of light energy had the same characteristics as particles, such as appearing at a single point in detectors, and they became known as photons.

Pion: the lightest *meson*, with mass of just over one-eighth of the *nucleon* mass.

Positron: the anti-particle of the *electron*. It has the same mass as the electron but with opposite (positive) electric charge. The beta radiation from nuclei consists of electrons or positrons together with the (very nearly) undetectable *neutrinos*.

Prolate deformation: the deformation of a sphere into the shape of an 'American football' shape. This can be achieved with a balloon, say, by pulling apart two points on either side of the balloon's surface.

Proton: one of the constituents of nuclei and the only constituent of the lightest nucleus, hydrogen. It has a positive charge of the same magnitude as the negative charge on an *electron*, and a mass almost two thousand times heavier than the electron. Every neutral atom contains as many protons in its nucleus as electrons orbiting outside it.

Quantum mechanics: the set of physical laws that govern the behaviour of the subatomic world. It began with the ideas of Max Planck and Albert Einstein at the beginning of the twentieth century and was developed into a complete mathematical theory by the mid-1920s by Neils Bohr, Erwin Schrödinger and Werner Heisenberg. Other physicists such as Paul Dirac, Max Born and Wolfgang Pauli made major contributions. While quantum mechanics is the most successful theory in science, underpinning much of modern physics, chemistry, electronics and material science, its predictions remain strange and counter-intuitive, especially when met for the first time.

Quark: component particle of *protons*, *neutrons*, *mesons* and other *hadrons*. They cannot exist isolated, outside hadrons. The proton consists of two 'up' quarks, each with a positive charge of magnitude 2/3 the charge of an electron, and one 'down' quark, with a negative charge 1/3 the charge of an electron; the neutron has two down quarks and one up quark, making it a neutral particle.

Quark–gluon plasma: when *nuclear matter* is raised to a high temperature and subjected to huge pressure, as happens in very high energy collisions between heavy nuclei, the boundaries between the *nucleons* disappear and the *quarks* and *gluons* within them form a sort of soup. The Universe very briefly passed through a quark–gluon phase before *protons* and *neutrons* emerged after the Big Bang.

Scintillation: the flash of light when a nuclear particle or *gamma ray* strikes certain substances.

Scintillation counter: a particle detector in which the light from *scintillations* is detected and measured with *photomultipliers*.

Segrè chart: a chart of the nuclei arranged (usually) with the *proton* number on the vertical axis and the *neutron* number on the horizontal axis. Generally, the square on this chart which corresponds to particular values of N and Z will indicate key properties of that particular nucleus, such as its radioactivity.

Silicon detector: modern detectors of charged particles are based on the fact that such particles cause electrical signals to be generated when they pass through or are absorbed. Such detectors allow the energies and directions of particles produced in nuclear reactions to be measured accurately.

Spallation: when a very high energy *proton* or other projectile strikes a nucleus, the nucleus is likely to shatter, producing an array of lighter nuclei. Nuclei such as ^{6}Li are believed to have been produced in spallation reactions in interstellar space.

Spectrometer: an instrument that separates radiation into its different *wavelengths* (or *frequencies*). Since every nucleus or every different atom radiates different sets of wavelengths, a spectrometer allows us to identify what atoms or nuclei are present in a sample. In addition, the particular pattern of wavelengths provides vital information about the atom or nucleus from which it has been radiated.

Spectrum: a representation showing how the strength (or brightness) of *electromagnetic radiation* from a given source depends on its *wavelength*. Also refers to the band of colours we see when light or other radiation is separated according to *frequency* (or wavelength); the most familiar example of this is the rainbow spectrum of visible light.

Spontaneous fission: the process whereby a heavy nucleus with an excess of neutrons splits roughly in half into two lighter nuclei. Such a process is a type of radioactive decay and takes place without the nucleus needing to absorb any surplus neutrons to prompt it to *fission* as would be needed in a nuclear reactor.

Super-heavy nuclei: nuclei with Z in the region of 110 or above.

Supernova, type Ia and type II: a supernova is the catastrophic explosion of a star in which it briefly emits as much radiation as all the stars in its galaxy. A type Ia supernova takes place when the companion to a white dwarf star sheds enough matter onto the white dwarf that it exceeds the maximum mass such a star can have. A type II supernova takes place when the nuclear fuel in a giant star is exhausted. Radiation pressure from nuclear reactions can no longer hold the star up, so it collapses inwards. It then bounces out as a result of the incompressibility of *nuclear matter*, releasing vast amounts of energy, a flood of *neutrinos* and much matter into space. Many elements of which we and the Solar System are made were produced in supernova explosions billions of years ago.

Synchrotron: developed from the *cyclotron* for much higher energies and involving a circular rather than spiral path. Synchrotrons can accelerate charged particles to very nearly the speed of light.

Theory of relativity: Einstein's special theory is based on two ideas: (i) that the speed of light in a vacuum (c) is always the same, no matter how fast you are moving relative to the source of light, and (ii) the laws of physics are the same no matter how fast your lab is moving at a steady speed. One consequence is that mass m and energy E are equivalent, and that $E = mc^2$.

Tokamak: an apparatus for producing nuclear *fusion* on Earth. It consists of a vacuum chamber in the form of a torus together with large magnets for keeping the interacting ions on closed paths through the vacuum.

Tunnelling (quantum): *quantum mechanics* permits particles to appear on the far side of barriers which, according to physics before quantum mechanics, they would not have the energy to surmount. This 'quantum tunnelling' controls such things as *alpha decay*, and permits the *fusion* of light nuclei in stars, allowing stars like the Sun to shine for billions of years.

Uncertainty Principle: one of the fundamental ideas in quantum theory, first expressed by the German physicist Werner Heisenberg, which states that for any object certain pairs of properties, such as position and momentum, cannot be known precisely at the same time. However, this feature is not a result of the inevitable 'clumsiness' of our measuring apparatus when it comes to subatomic objects, but rather is inherent in the objects themselves.

Wavelength: the distance between two consecutive crests (or troughs) of a wave; wavelength is inversely proportional to the frequency of the wave. For electromagnetic radiation, the wavelength is equal to the speed of light divided by the frequency.

Wave–particle duality: the quantum concept whereby both matter and radiation must be regarded at their most fundamental level as sometimes having wave properties and sometimes particle properties. For example, *electrons* and *photons* sometimes behave like particles and sometimes like waves.

X-ray: a form of electromagnetic radiation with *wavelengths* shorter than ultraviolet but longer than *gamma rays*. Since the wavelength of the radiation is related to energy this means that X-ray photons are less energetic than gamma ray photons. However, there is no sharp dividing line between the two. X-rays have wavelengths ranging from about 10 nanometres down to 10 picometers (see table on page 15).

Subject index

Bold type indicates an illustration